CW00486644

Institut Pasteur

TODAY'S RESEARCH
TOMORROW'S MEDICINE

“Scientific research is one of the most rewarding and inspiring of all human enterprises, especially when the aim is to tackle disease and pain, to alleviate human suffering. The Institut Pasteur has tirelessly pursued this goal for the past century, and as it embarks on the next hundred years it is more determined than ever to persevere in its mission.”

François Jacob, Chairman of the Institut Pasteur
Board of Directors (1982-1988),
at the opening ceremony for the centenary
of the Institut Pasteur on October 5, 1987.

Institut Pasteur

TODAY'S RESEARCH
TOMORROW'S MEDICINE

Forewords by Christian Vigouroux and Érik Orsenna

ABRAMS | NEW YORK

Contents

Overcoming the impossible

The Institut Pasteur has always played a vital role in society, from its beginnings to the present day, and it will no doubt continue to do so in the future. Its passionate dedication, audacious scientific spirit and global vision have placed it firmly on the front line in battling the major scourges that plague human health. The Institut Pasteur is constantly reinventing itself, challenging certainties and shifting its focus to keep pace with new discoveries—both its own and those of others. It is a place where the impossible becomes possible, where the unknown can become known. That inspires confidence and recognition; it raises hopes and expectations.

It can be hard to fully grasp just how much the Institut Pasteur has contributed to humanity, but the article on "Rabies" in Bouillet's *Dictionnaire universel des sciences, des lettres et des arts* can give us an idea. We read of *"a virulent disease characterized by severe disruption to the nervous system that affects sensitivity, movement and intelligence. The only remedy is immediate cauterization with a red-hot iron. Caution is advised when considering any recommended treatments, as they have no real effect."* That was in 1884—nothing was possible. But the following year, the article had to be rewritten—suddenly, everything had become possible.

Since the time of its founder, that has been the role of the Institut Pasteur: disproving evidence, daring to tread the risky path of theory and exploration rather than relying on received knowledge, exchanging ideas with experts all over the world and, against all the odds, pursuing its core missions of research, education and public health.

When I arrived at the Institut Pasteur, I was struck by a series of qualities that encapsulate and justify the Pasteur "brand". For "Pasteur" opens doors and elicits recognition, in every sense of the word, because it is committed to inventing, discovering, saving lives.

Courage. Scientists are plagued by self-doubt, failure and the inevitable tension between ambition and reality. They are expected to wear two hats—they are both inventor and banker, developing both research agendas and funding plans that ultimately serve the same aim.

Curiosity and risk-taking. Scientists need to be able to define a problem, to weigh it up against the calculations and findings of other scientists, to dare to go out on a limb. They also need to be able to celebrate their failures, recognizing them as stepping stones to future success.

A global approach. Braudel coined the term "world economy", and one might say that the Institut Pasteur is engaged in "world research". It works with and alongside the world's major research institutes, striving for constant improvement through an effective process of benchmarking. The Institut Pasteur is unique in its global, universal approach to human health. With 33 institutes in 26 countries on every continent, the Institut Pasteur is sowing seeds in every corner of the planet to heal the ills of humankind.

A sense of cooperation and teamwork. Researchers are always in competition—few other professions are evaluated on a constant basis by independent international committees. But despite this pressure (or maybe because of it), researchers work together in communities, in tightly-knit teams. They are part of a research unit, a department, a group. Researchers are defined by their team, and increasingly also by the links their team forges with other research communities.

A grounding in reality, a desire to understand patients and not just their diseases. The Institut Pasteur is involved in *translational* research, carried out in close cooperation with health care workers. This focus on reality also means keeping a constant eye on diseases, making sure National Research Centers—which have the task of identifying the emergence of risks and raising the alert at national and international level—can do their job effectively.

Finally, a spirit of cooperation and communication. Since its early days, the Institut Pasteur has been committed to teaching and transferring the latest knowledge. But more than that, it has fostered a unique collaborative model, based on the mutually beneficial interaction and intermingling of its three core missions.

The Institut Pasteur is a force for global health, for the relief and assistance of populations, through the pursuit of progress. It is always striving to tackle major infectious diseases, always on guard—and nowadays it even wages its battle at cellular level, thanks to advances in molecular biology and neuroscience. We know only too well that scourges of the past are capable of coming back to plague us in new forms in the future. But the Institut Pasteur is ever vigilant, constantly keeping watch and doing its utmost to safeguard our future.

I sincerely hope that this book will raise awareness of the tremendous potential of the Institut Pasteur so that it may continue its work for the benefit of us all.

Christian Vigouroux
Chairman of the Board of Directors of the Institut Pasteur

Life, death, life

When I unwittingly found myself occupying Louis Pasteur's seat at the Académie française, I felt that I should find out more about my illustrious predecessor. François Jacob asked me to write a biography. Trembling with nerves, I set about my task, working fervently and passionately. Once I had finished, I needed a title that would sum up what was a remarkably full, generous existence.

La vie—life—seemed the only possibility.

Has any other genius taken greater strides to advance knowledge? From his early work on fermentation to his vaccine for rabies, Pasteur was able to make the leap from magic to science, from merely describing living phenomena to understanding them, from passive hopelessness at the existence of so many devastating diseases to genuine hope that they might be treated.

But I needed to add another word to my title: *la mort—death*.

For Louis Pasteur, death was ever present; it was relentless. It claimed three of his beloved daughters, taking its revenge because his discoveries had stripped it of some of its power. It threatened to take him prematurely too, afflicting him with a first, severe stroke at the age of just 46—but it was unable to slow him down. Death was his one enemy, his lifelong combat.

But the cycle continued. *Life* prevailed.

For who can believe today that Pasteur has really gone?

Through his institute, the center of a vast network covering 26 countries, his noble ambition lives on: to advance understanding, care for those in need and share knowledge.

Life, death—but always followed by life.

Erik Orsenna
Writer, Member of the Académie française
Ambassador for the Institut Pasteur

The Beginnings of the Institut Pasteur

INTERVIEW WITH MAXIME SCHWARTZ AND MICHEL MORANGE

Accounts from Maxime Schwartz, President of the Institut Pasteur from 1988 to 1999 and author of several books on the history of the institute and its stakeholders, and Michel Morange, Director of the Cavaillès Center for History and Philosophy of Science at the École Normale Supérieure in Paris, and specialist in the history of life sciences in the 20th century.

Today, with its 2,500-strong staff and worldwide network of 33 institutes, the Institut Pasteur seems far removed from the historical building established on market gardening land in the Vaugirard district of Paris in 1888. However, its pioneers instilled a scientific spirit and approach that are every bit as strong today.

How did the Institut Pasteur come about?

Maxime Schwartz: Throughout his career, Louis Pasteur fought to get premises where he could conduct his experiments. But it was only after vaccinating little Joseph Meister against rabies in July 1885, and the young shepherd Jean-Baptiste Jupille in October of the same year, when people who had been bitten in France and the world over started flooding to the École Normale where Pasteur worked, that he came up with the idea of an institution dedicated to these vaccinations. On March 1, 1886, the French Academy of Sciences, which had been won over by the idea, launched an international appeal to fund the project. It was so successful that Pasteur was able to broaden his goals. His institute would obviously treat bitten people, but it would also be a center for education and research into infectious diseases.

Opposite: Oil on canvas by Albert Edelfelt (1886) showing Louis Pasteur in his laboratory at the École Normale Supérieure in 1885.

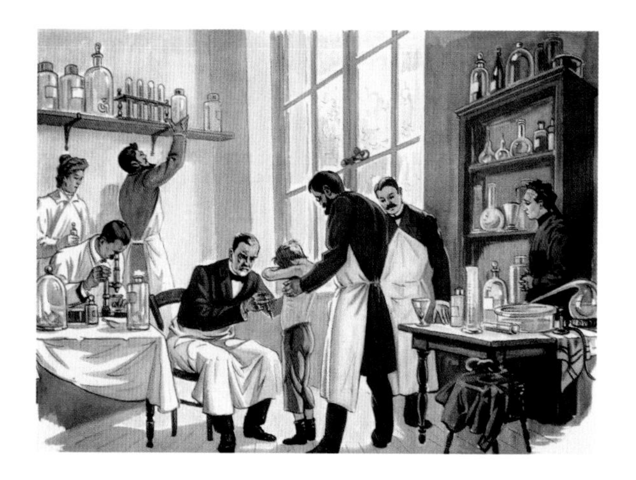

THE RABIES VACCINE, PAST AND PRESENT

When Louis Pasteur began studying rabies in 1880, the disease was dreaded by all. Even though it claimed fewer lives than tuberculosis or cholera, those infected died a painful death. Today, thanks to the rabies vaccine he developed in 1885, nobody dies of rabies... provided that they are treated in time and a specific protocol is followed (wound cleansing, repeated vaccine injections and sometimes serum therapy) but this is far from being the case everywhere. Although no cases of human rabies acquired in mainland France have been reported since 1924, the disease still causes several tens of thousands of deaths throughout the world each year, mainly in Africa and Asia and most often following a bite from a rabid dog.
There are many reasons for this, for instance dogs are not vaccinated, treatment is too costly and rabies vaccination centers are rare in some countries. The Institut Pasteur has fought this disease from its very beginnings. To start with, it vaccinated the hundreds of infected people that arrived at its door. Today, the National Reference Center for Rabies, directed by Hervé Bourhy, Laurent Dacheux and Perrine Parize, at the Institut Pasteur in Paris, carries out epidemiological monitoring of the disease in France and leads a vast network of anti-rabies treatment centers.
Several International Network institutes are also involved in local anti-rabies initiatives. In Cambodia, particularly, where 600,000 people are bitten each year and where rabies is responsible for roughly 800 deaths annually, the local Institut Pasteur vaccinates over 21,000 people but most bite victims do not have the means to get to the center. This institute is also carrying out research to reduce the number of vaccine doses and lower the cost of treatment. A government plan is also being drawn up to boost monitoring and set up frontline rabies prevention centers.
Further research, conducted at the Institut Pasteur in Paris, aims to shed new light on the rabies virus. Following an infected bite, the virus rapidly reaches the muscle nerve endings then it travels to the spinal cord before spreading to the brain where it causes encephalitis and, ultimately, death. The Viral Neuro-Immunology Unit, directed by Monique Lafon, is studying how the virus evades the human immune system and forces the infected neurons to survive. The aim is to develop new molecules for treating neurodegenerative diseases and repairing nerve and spinal cord damage. As for the Antiviral Strategies Unit, led by Noël Tordo, it is investigating antiviral molecules capable of blocking the virus when it has already reached the nervous system (we are currently unable to cure rabies once the nervous system is affected). Research is underway into molecules with strong antiviral activity. Finally, we know that each animal species (fox, dog, bat, etc.) is infected by specific types of the virus responsible for rabies—lyssaviruses. The Lyssavirus Dynamics and Host Adaptation Unit and WHO Collaborating Center for Rabies are studying these viruses, under Hervé Bourhy, to answer a whole host of questions. What is their evolutionary history? Why is the fox strain less pathogenic than the dog one? Bat lyssaviruses do not spread easily to humans but could this situation change? One day, the team hopes to block certain lyssavirus protein functions and produce new treatments, and ultimately to improve understanding of the mechanisms underlying the emergence of RNA viruses originating from animal reservoirs. The rabies story is far from over...

Which infectious diseases did Pasteur and his team study first?

M.S.: To begin with, Pasteur mainly focused on animal diseases. He was not a doctor and did not dare tackle human diseases. It was also much easier to conduct experiments on animals rather than humans. He developed vaccines against fowl cholera (1879) and anthrax (1881), a disease that killed 20 to 30% of cattle and sheep in some regions. Rabies gave him the chance to study a disease that affected both animals and humans. From 1888 onwards, the most significant research concerned diphtheria and was led by Émile Roux and Alexandre Yersin, two of his "lieutenants".

Why diphtheria?

M.S.: Roux was 10 years older than Yersin and had been involved in many of Pasteur's research projects. Yersin was the newcomer, today's equivalent of an intern. He assisted both Jacques-Joseph Grancher, a pediatrician at Necker Children's hospital, and Roux, first at the École Normale then at the Institut Pasteur. The two men were interested in diphtheria for two reasons. Firstly, Yersin witnessed the disease—which at the time killed several tens of thousands of children every year in France—first hand while working at the Children's hospital. It was awful to watch diphtheria take hold, as the children suffered from increasingly frequent and violent attacks of suffocation until they died. Yersin was desperate to find a solution to this dreadful fate.

As for Roux, he could see the benefit in terms of science. Along with Charles Chamberland, another man on Pasteur's team, he had demonstrated that it was possible to immunize animals against bacteria that causes septicemia. In culture, these bacteria secreted a substance that could be used to vaccinate animals against the disease. Roux thought that perhaps the diphtheria bacillus was like these bacteria.

And it was...

M.S.: Yersin and Roux showed that the bacterium responsible for diphtheria secretes a toxin—a poison that is behind the main symptoms of the disease—and they believed they could develop a diphtheria vaccine using

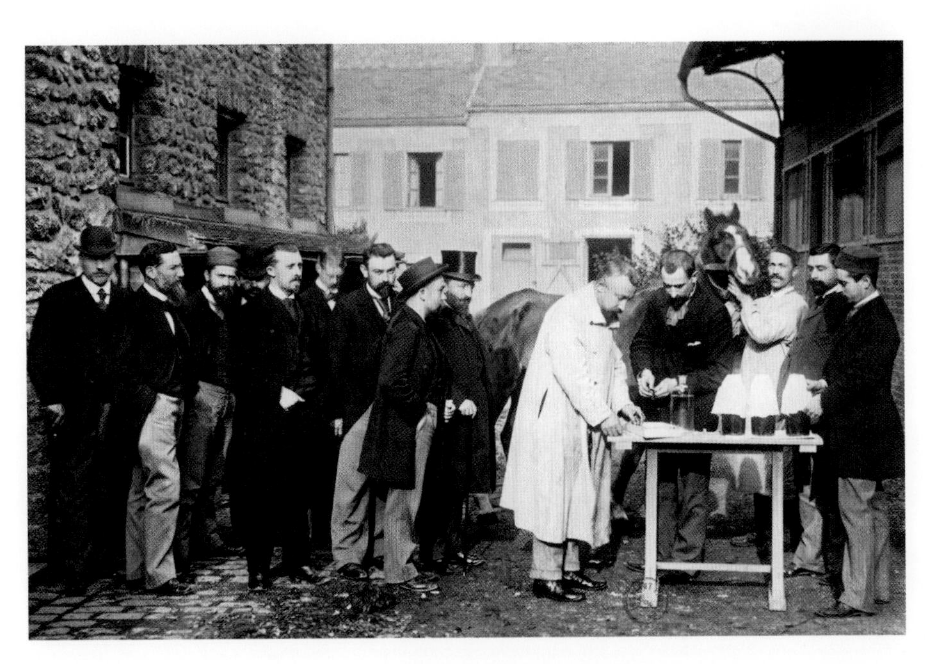

Opposite: Artist's impression of Louis Pasteur vaccinating Joseph Meister or another child against rabies.

Above: Blood sample taken from a horse immunized against the diphtheria toxin (1900). The use of horse serum (serum therapy) led to a steep drop in the mortality rate for infected children.

LOUIS PASTEUR, A DETERMINED VISIONARY

"Adopt a critical mind. [...] What I am asking you now and what you will, in turn, ask your disciples is the most difficult part of being an inventor. Believing that you have made a scientific breakthrough, desperately wanting to make it public but forcing yourself to wait days, weeks and sometimes even years, trying hard to ruin your own experiments and only announcing your discovery when you have exhausted all other possibilities, yes, it's a laborious task. But when, after so much effort, you are finally certain, you experience one of the greatest joys ever felt by the human soul."
When his institute opened on November 14, 1888, Louis Pasteur (1822-1895) was not well enough to give the inaugural speech. But his resolve shone through in these words, read out by his son Jean-Baptiste before the French President Sadi Carnot and other prominent figures. Pasteur showed proof of determination from an early age when, while still at school, he set his sights on the École Normale in Paris and overcame the hurdles to get there. He also approached all the scientific questions that guided his research with the same enthusiasm and tenacity, whether it was to understand research into the effects that some crystals have on light during his PhD thesis, to investigate the cause of fermentation in Lille, to prove that the spontaneous generation theory championed by some was nothing more than a microbial contamination problem or to highlight the microbial origin of the various diseases he encountered during his career—those that decimated silk farms in the south of France in the 1860s, those specific to grapes and hops and also those that affected animals, like anthrax, or humans, such as nosocomial infections or rabies. And he was rewarded for his determination every time. Not only did he find the causes of these diseases but his solutions for fighting them were effective and are still vital today, for example pasteurization, sterilization (of dressings and surgical instruments) and vaccination.

LOUIS PASTEUR'S FIRST "LIEUTENANTS"

ÉMILE DUCLAUX

Pasteur's very first "lieutenant" Émile Duclaux (1840-1904) was also committed to bringing science closer to its applications. After research into silk worm and grape diseases with Pasteur, he focused on cheese. Both a physicist and chemist, he became one of the pillars of microbiology thanks to his work on microbes.

ÉMILE ROUX

The third director of the Institut Pasteur after Louis Pasteur and Émile Duclaux, Émile Roux (1853-1933) devoted his life to science. Despite his poor health, he successfully managed his research on diphtheria and tetanus, his courses and the institute itself.

ALEXANDRE YERSIN

In 1894, doctor and microbiologist Alexandre Yersin (1863-1943) was sent to Hong Kong by the French government and the Institut Pasteur to investigate an outbreak of the plague. He discovered the bacillus responsible for the disease in just a few days and today it is named *Yersinia pestis*. A few years later in 1898, Paul-Louis Simond, another Institut Pasteur scientist, highlighted the role of fleas in spreading the plague from rats to humans.

ALBERT CALMETTE

As a naval physician, Albert Calmette (1863-1933) was confronted with various diseases, such as yellow fever, during his postings to Hong Kong, Saigon and Gabon before he joined Pasteur's team in 1890. He set up the first institute outside France in Saigon in 1891, then the one in Lille in 1895. Albert Calmette, together with Camille Guérin, developed the tuberculosis vaccine, or BCG.

this product. But the Germans were quicker off the mark. As early as 1890, in the institute headed by Robert Koch in Berlin, Emil von Behring and a Japanese scientist working in his lab, Shibasaburo Kitasato, showed that when partially inactivated, this toxin could be used to immunize animals against diphtheria and, better still, an antitoxin (now known as an antibody) was found in the blood of immunized animals. They therefore came up with the idea of treating children with diphtheria using blood from immunized animals containing the antitoxin. This marked the start of serotherapy, which was developed by Behring in Germany and Roux and his team at the Institut Pasteur at the same time. It helped to save a great many children from diphtheria.

During the Great War and thanks again to Behring and Roux, serotherapy would also protect thousands of soldiers against tetanus—a disease which, like diphtheria, is caused by toxin-secreting bacteria.

Pasteur was surrounded by many people in the 1880s. Who was in his inner circle and what role did they play in founding the institute?

M.S.: When the institute opened in November 1888, Pasteur was suffering from ill health. He was already elderly and had had several strokes. So, even though he was actively involved in the initial project and in drawing up the articles of association—which ensured that the institute remained independent from public authorities and industry—, it was Émile Duclaux and Émile Roux who saw the project through to completion. A Russian scientist, Ilya Mechnikov, made a significant contribution to research at the institute, and Yersin and Albert Calmette, a naval physician, were instrumental in setting up the international network (see page 36).

THE INSTITUT PASTEUR AND THE BIRTH OF IMMUNOLOGY WITH ILYA MECHNIKOV

When he joined the Institut Pasteur in December 1888, Ilya Mechnikov brought with him an emerging scientific discipline—immunology. Born in Ukraine in 1845, this zoologist began by studying embryology and marine biology. In 1882, he observed the reaction of transparent starfish larvae after introduction of a rose thorn—the thorn was surrounded and engulfed by motile cells. The discovery of these cells, which he named phagocytes (from the Greek *phagein*, meaning to eat and *kytos*, cell), laid the foundations for cell immunity. A committed Darwinian, Mechnikov believed that phagocytosis—the process by which cells ingest or engulf particles—was a way for living organisms to defend themselves against intruders and also microbes. He then sought to show that this was a key factor in inflammation and pus formation.

As an avid scientist, who was both committed and convincing, Mechnikov was involved in a controversy with Emil von Behring and the German school, who were proponents of a humoral immunity theory founded on the bactericidal effects of serum. We know today that both hypotheses are correct, that they complement each other and are not mutually exclusive. In recognition of his work, Mechnikov received the Nobel Prize in Medicine in 1908, alongside Paul Ehrlich who described the toxin-antitoxin reaction and, more broadly, the antigen-antibody interaction—cell and humoral immunity were therefore both recognized by the Swedish jury.

Jules Bordet, a Belgian Institut Pasteur scientist and pupil of Ilya Mechnikov, highlighted the role of alexin (or complement) in the destruction of bacteria by antibodies. This work earned him the Nobel Prize in Medicine in 1919. As well as his scientific vision and tireless work ethic, Ilya Mechnikov—who Émile Roux said "looked just like a science demon", was a gifted teacher and trained a generation of talented researchers.

What contributions were made by Duclaux and Roux?

M.S.: At the top of the steps to the historical Institut Pasteur building, visitors are greeted by two busts—Duclaux and Roux. Duclaux was one of Pasteur's very first "lieutenants". He set out the organization for the institute while Roux was involved in its design and construction. When Pasteur died in 1895, Duclaux officially succeeded him but he was already at the helm of the institute before his mentor's death. After Duclaux died in 1904, Roux took over and remained director for almost 30 years. The institute was his life. He was a man of action who was always very present. He instilled a certain spirit that lives on today, "the Pasteurian spirit".

How did a Russian scientist become part of the management team?

M.S.: Mechnikov had visited Pasteur in October 1887 and, struck by the size of the institute under construction, asked if he could work there. Pasteur, who was very interested in his research on the body's fight against microbes, not only invited him to work there but gave him access to a laboratory and then made him head of a department. From a scientific point of view, Mechnikov was indisputably a leader. He attracted huge numbers of French and Russian students, and everyone turned to him for scientific advice. His work on immunity, which earned him the Nobel Prize in 1908, shaped immunology and microbiology in the first half of the 20th century.

How would you describe the Pasteurian spirit?

M.S.: An important aspect is the link between research and its applications. Pasteur's own career was marked by constant toing and froing between academic and clinical research. For example, his initial, academic research led him to believe that asymmetric molecules, like the tartrates he had studied, were only produced by living beings. So, when he studied fermentation because the father of one of his students in Lille was struggling to produce alcohol from beetroot juice, he was struck by the fact that asymmetric molecules were produced during fermentation and concluded that they were caused by microorganisms. This led, firstly, to applied research involving vinegar and wine and then to more

Portrait of Ilya Mechnikov (1845-1916) around 1910. After discovering phagocytes and phagocytosis in 1882, Mechnikov developed a pioneering theory on cellular immunity and went on to win the Nobel Prize in 1908.

academic research that aimed to show that these microbes do not appear spontaneously, therefore disproving the spontaneous generation theory.

Another aspect is humanism—the interest in diseases in the least developed countries, malaria for example. Research into these diseases is not of any direct interest to our fellow citizens nor does it bring obvious financial benefit, but Institut Pasteur scientists devoted a significant amount of time to their study.

Michel Morange: There was also the belief that knowledge from the Institut Pasteur should be exported to serve all countries and this was a fairly unique view for the time. The best illustration of this was the Institut Pasteur course introduced by Roux and Yersin in 1889. Scientists came from all over the world to study microbiology at the institute.

What were the greatest triumphs of the Institut Pasteur in the early days?

M.S.: In 1894, the same year that serotherapy was developed, Yersin discovered the plague bacillus in Hong Kong and prepared an anti-plague serum. Later, in the 1920s, Institut Pasteur veterinarian Gaston Ramon observed that if diphtheria and tetanus toxins were treated with formaldehyde, they became totally inactivated but did not lose their immunizing power. Therefore, unlike partially inactivated toxins used to obtain antitoxic serum in horses, they could be used to directly inoculate children as a preventive measure. These children would then produce protective antibodies themselves without having to receive animal serum. This signaled the discovery of the diphtheria and tetanus vaccines, which replaced serotherapy and are still in use today. There is consequently no more diphtheria in France. The bacteria responsible for the disease have not however been eradicated. In fact, there was an outbreak in the USSR in the 1990s due to inadequate vaccination coverage. This is why it is vital to continue vaccination against diphtheria and tetanus.

Another significant discovery was the development of the tuberculosis vaccine, or BCG, by Calmette and Camille Guérin (see page 46). Jacques Tréfouël, Daniel Bovet and their team also discovered the anti-infectious action of sulfonamides in the 1930s. These molecules were used to treat numerous bacterial diseases, including leprosy.

What about academic research?

M.S.: From its foundation until the 1960s, the Institut Pasteur was very focused on medical issues. Then along came molecular biology, partly thanks to André Lwoff, François Jacob and Jacques Monod—the three Institut Pasteur scientists who were awarded the Nobel Prize in Physiology or Medicine in 1965—and their team. More academic research therefore followed. Subsequently, when Monod became President of the Institut

Above: Plate by Lackerbauer. Illustration representing microbes observed by Pasteur during his work on fermentation.

Opposite: Page from Louis Pasteur's laboratory notebook on butyric fermentation, dated February 12, 1861.

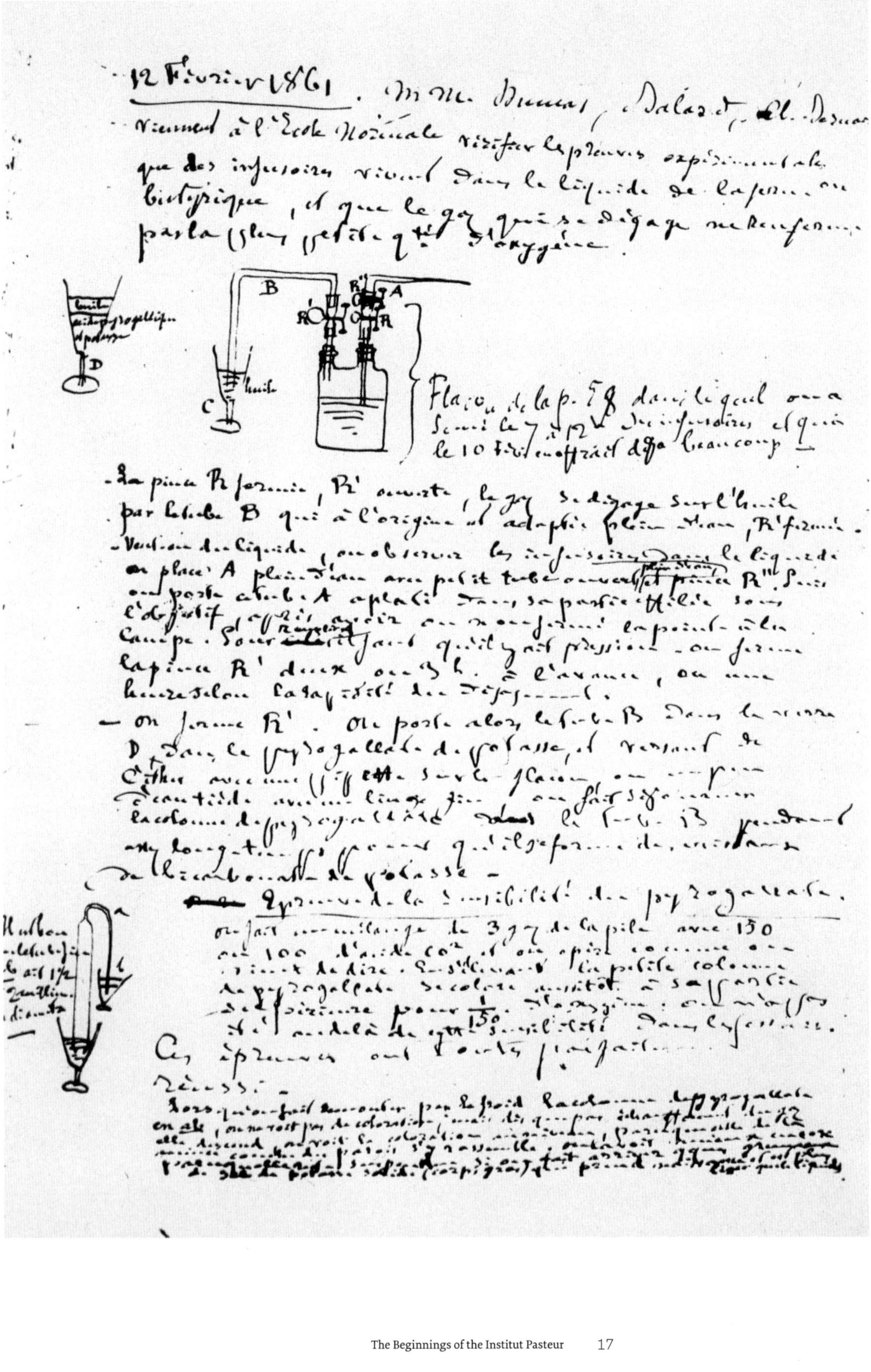

12 Février 1861. MM. Dumas, Balard, Cl. Bernard viennent à l'École Normale vérifier les preuves expérimentales que des infusoires vivent dans le liquide de la fermentation butyrique, et que le gaz qui se dégage ne renferme pas la plus petite qté d'oxygène.

huile
acide pyrogallique
et potasse
D

B R' A R R C huile

Flacon de la p. [illegible] dans lequel on a semé le 7 à 12h des infusoires et qui le 10 [illegible] offrait déjà beaucoup

- La pince R fermée, R' ouverte, le gaz se dégage sur l'huile par le tube B qui à l'origine est adapté plein d'eau, R' fermé.
- Veut-on du liquide, ou observer les infusoires dans le liquide on place A plein d'eau avec petit tube ouvert et pince R''. Puis on porte ce tube A aplati dans sa partie effilée sous l'objectif après avoir [illegible] la pointe à la lampe. Pour qu'il y ait pression on ferme la pince R' deux ou 3 h. à l'avance, ou une heure selon la rapidité du dégagement.
- On ferme R'. On porte alors le tube B dans la cuve D dans le pyrogallate de potasse, et versant de l'éther avec une pipette sur le flacon, ou [illegible] d'eau tiède avec un linge fin, on fait se former la colonne de pyrogallate dans le tube B pendant assez longtemps [illegible] qu'il se forme du [illegible] de bicarbonate de potasse –

Épreuve de la sensibilité du pyrogallate.

On fait un mélange de 3 [illegible] de la pile avec 150 cc 100 d'acide CO^2 et on opère comme on vient de dire. En s'élevant la petite colonne de pyrogallate se colore aussitôt. [illegible] pour $\frac{1}{150}$ d'oxygène; [illegible] au delà de cette sensibilité dans les essais.

Ballon [illegible] 1 1/2 [illegible]

Ces épreuves ont toutes parfaitement réussi –

Lorsqu'on fait remonter par le froid la colonne de pyrogallate en elle, on ne voit pas de coloration, mais dès que par échauffement [illegible] elle descend on voit la coloration [illegible]

Pasteur in 1971, he wanted the concepts and techniques of molecular biology to be used for research into infectious microorganisms. Researchers like Philippe Sansonetti were among the first to go down this road.

Back then it was complicated because genetic engineering was not yet available. But, when I took up my post as President of the Institut Pasteur, progress had been made and one of our main aims was to use molecular biology to tackle microbial issues. There was also new urgency in the medical field because several infectious diseases had emerged, AIDS in particular, when it was thought that they were under control thanks to antibiotics and vaccines. The molecular approach was gradually adopted to study these diseases and is today used throughout the world.

How did Institut Pasteur scientists progress from studying microbes to molecular mechanisms?

M.S.: There was not, strictly speaking, a shift from microbe to molecule. Lwoff, Jacob and Monod began by studying microbes and the molecule was the natural next step.

M.M.: Working on microbes was an advantage for the Institut Pasteur, as the molecular revolution occurred through research into bacteria and bacterial viruses—bacteriophages. These organisms are simpler than so-called higher organisms and are better suited to what is known as "molecular reduction".

M.S.: In fact, molecules actually appeared in biology in the 1960s with the discovery of the structure of DNA (1953), the molecule that carries hereditary genetic information in cells. Before this, Monod focused on bacterial culture problems, and Lwoff and Jacob studied bacteriophages, which were discovered exactly 100 years ago by another Institut Pasteur scientist, Félix d'Hérelle.

What led them to molecules?

M.M.: For Lwoff, two lines of research emerged in the 1950s. The first one sought to understand a curious phenomenon—in the Sorbonne laboratory where he was an assistant in 1940, Monod had observed that if two different sugars (glucose and lactose) were added to a bacterial culture at the same time, the bacteria only fed on the second sugar when they had finished the first one. This was something that Pasteur and then Duclaux had also noticed in their time. Lwoff thought that this behavior was due to the appearance of enzymes in the bacteria that were capable of breaking down the lactose once the glucose had run out. This became Monod's topic, first in Lwoff's team at the Institut Pasteur, then in the laboratory that he led there.

The second theme focused on a phenomenon that was specific to the life of bacteriophages—some of them infected bacteria then remained dormant for generations until they "woke up", multiplied in the bacteria, released themselves and destroyed them. This phenomenon was long seen as an artifact but, in the 1930s, Eugène and Elizabeth Wollman, two brilliant biologists from the Institut Pasteur, lent support to the idea that the concerned bacteria retained the bacteriophage, probably in an inactive form, and Lwoff came up with a term for this—the prophage. Understanding how the prophage is inhibited then reawakened became a key question and Lwoff assigned Jacob, then a young doctor, to the task as he wanted to focus on research.

M.S.: In both cases, there is this notion of genetic material that is inactive most of the time but, at some point, wakes up. In the first case, there are genes that only produce enzymes if an inducer is added. In the second,

Opposite: The three laureates of the 1965 Nobel Prize in Medicine—François Jacob, Jacques Monod and André Lwoff—on October 16, 1965.

LWOFF, MONOD AND JACOB—THREE INSTITUT PASTEUR SCIENTISTS BEHIND THE DEVELOPMENT OF MOLECULAR BIOLOGY

In 1965, the Nobel Prize in Physiology or Medicine was awarded to André Lwoff, Jacques Monod and François Jacob "for their discoveries concerning genetic control of enzyme and virus synthesis". These discoveries can be summed up in one word, operon—the name given to the gene activity regulation system that the three biologists from the Institut Pasteur and their team revealed in the "attic", the hundred or so square meters under the rafters of the Duclaux building that they used as a laboratory.

In the late 1950s, scientists had recently discovered that the DNA molecule in a cell carries genetic information that is passed down from one generation to the next. It was also understood that genes play an essential role in determining the characteristics of living organisms. But one question remained: how could cells carrying the same genes have different functions? Lwoff, Jacob and Monod solved the mystery by introducing the notion of regulatory genes. By comparing two observed phenomena—one in bacterial cultures, the other during infection of bacteria by viruses (bacteriophages)—the biologists showed that the genes of one cell are not always active. Their activity depends on other genes, called regulators, which encode proteins, which in turn inhibit, or conversely, activate groups of genes involved in the same function, sugar digestion for example. An operon is the term used by Jacob and Monod to describe a set of genes subject to the same regulation. This discovery helped to revolutionize biology by bringing it into the molecular era.

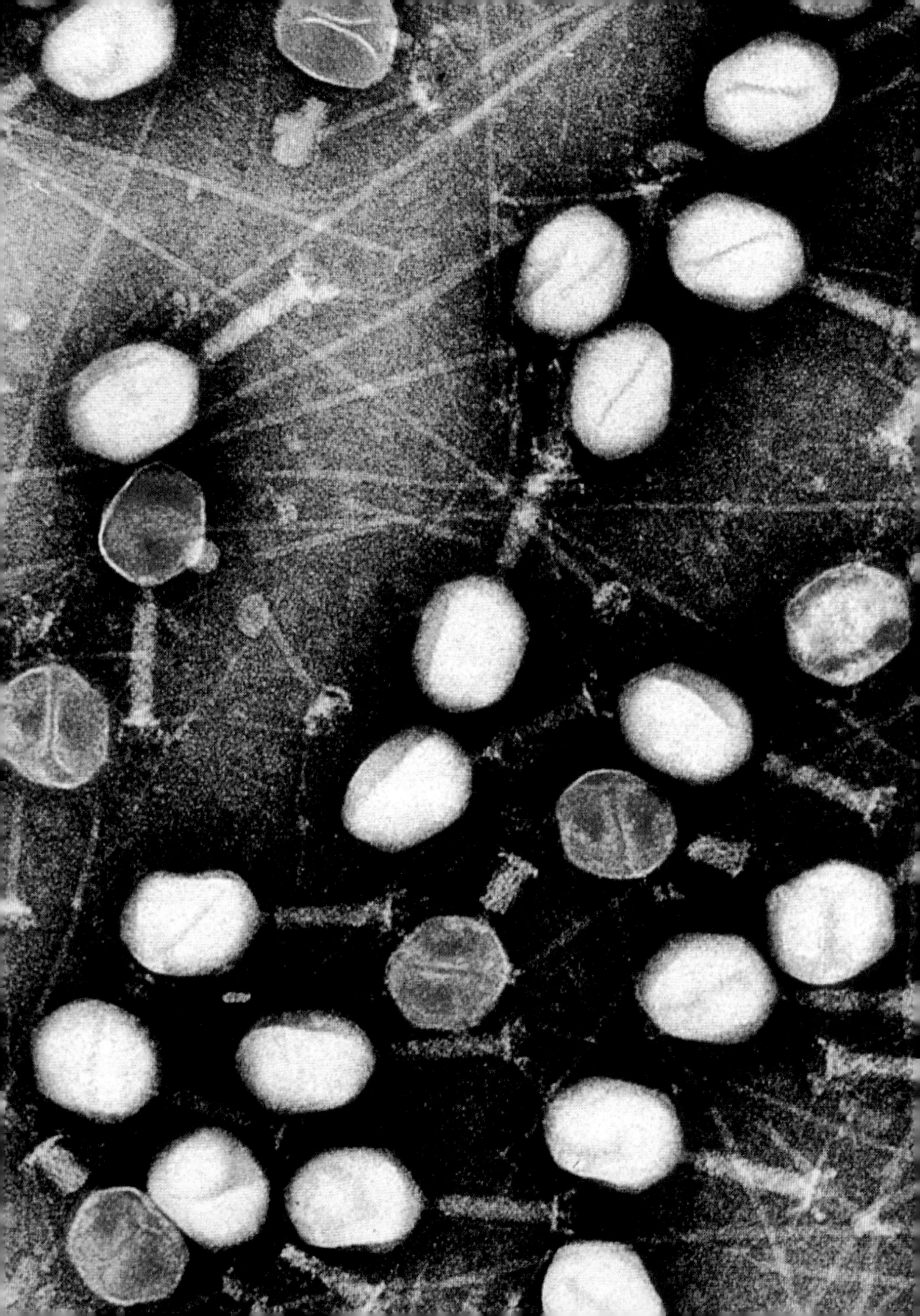

the prophage only wakes up if certain conditions are fulfilled. Research bringing these two phenomena together won the Nobel Prize in 1965 (see page 19).

M.M.: As early as the 1930s, Lwoff had an inkling that a shared mechanism linked the two topics and it was always at the back of his mind. It was through experiments conducted by Jacob and Monod to understand this mechanism that the team was gradually able to describe it at molecular level.

What impact did this discovery have?

M.S.: It was one of the two main pillars of molecular biology. The first being the demonstration that DNA carries genetic information and—thanks to the DNA structure discovered by James Watson and Francis Crick in 1953—that this genetic information is passed down through the generations. Institut Pasteur scientists then made another major breakthrough when they shed light on how genes produce proteins and how this production can be regulated.

M.M.: This signaled the advent of molecular biology. And, with the arrival of genetic engineering and the necessary tools a few years later, it changed the face of microbiology and biology.

M.S.: This was how in 1985, two to three years after the arrival of genetic engineering in France, Pierre Tiollais, an Institut Pasteur doctor and biologist, developed the hepatitis B vaccine—the first human vaccine obtained from animal cells using this technique.

How did this revolution in biology influence research at the Institut Pasteur?

M.M.: The work of Lwoff, Jacob, Monod and their team on the regulation of gene expression had a huge impact on research into embryogenesis. At the time, researchers had come up against a problem that the American biologist Thomas Morgan explained as follows in his Nobel speech in 1933—all cells in complex higher organisms, such as human beings or mice, have the same genes, but they are very different, probably because their genes are not

HOW DID NEUROSCIENCE REACH THE INSTITUT PASTEUR?

It was thanks to Lwoff, Jacob and Monod's research on gene activity regulation that neuroscience arrived at the Institut Pasteur, at the behest of Jean-Pierre Changeux. Between 1959 and 1963, Monod asked himself how a protein, whose role is to inhibit gene activity, could in turn be blocked. More generally, together with Jean-Pierre Changeux, who was at the time a PhD student under his supervision, he focused on the mechanisms involved in protein activation and inhibition. Monod and Changeux developed the concept of "allosteric" proteins, which can take on different forms depending on whether an inhibitor or activator is bound to them. Research into allosteric proteins became very popular in the 1960s.
This was when Jean-Pierre Changeux, who had returned to the Institut Pasteur following postdoctoral research in the US, questioned whether or not some proteins involved in intercellular communication were also allosteric. At the junction between neurons and muscles, a small molecule secreted by the nerve ending, acetylcholine, does indeed activate the muscle cell by binding to a receptor. By studying this mechanism in *Torpedo* or electric rays, Jean-Pierre Changeux demonstrated that this receptor is allosteric—when the acetylcholine binds to the receptor, it forms a pore in the cell membrane that ions then cross and which leads to muscle excitation.
Drawing on this discovery, Changeux was able to set up a new unit at the institute in 1972—the Molecular Neurobiology Unit. Neuroscience emerged at the Institut Pasteur and has never left.

Opposite: T2 bacteriophages, natural "predators" of *Escherichia coli* bacteria.

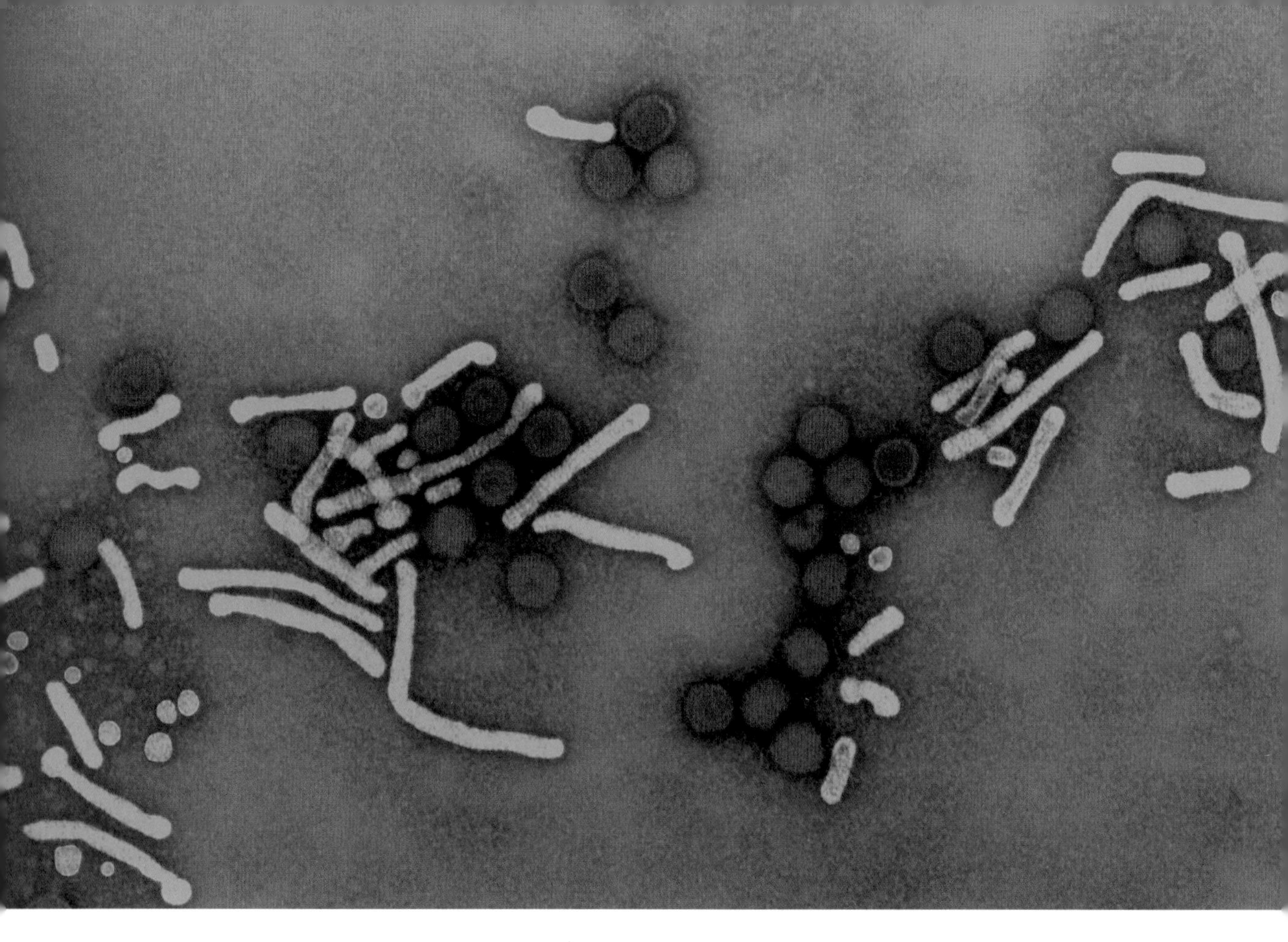

active in the same way. To understand their development, we need to understand how the activity of these genes is controlled. However, back then, nobody knew how genes were controlled or even what genes were!

The Institut Pasteur discovery did not provide a clear answer but it did give an idea of what developmental regulatory mechanisms could be, and provided the molecular tools for their identification. In the 1960s, several molecular biologists focused their research on the development of various organisms, such as the fly for Seymour Benzer, the nematode worm for Sydney Brenner and the mouse for Institut Pasteur scientists François Jacob and François Gros.

It is also thanks to the work of Jacob and Monod that neuroscience arrived at the Institut Pasteur (see page 21).

The hepatitis B virus viewed using transmission electron microscopy. In red, infectious viral particles (Dane particles); in green, non-infectious particles.

What remains of the early Institut Pasteur today?

M.S.: Without a doubt, a very outward-looking approach. A great many foreign nationals come to work at the institute and vice versa. The link with research applications is also still very strong and this is crucial, as patent royalties enable the institute to have its own funds alongside public funding and donations, as was Pasteur's wish. There is also the Institut Pasteur International Network, which has never stopped expanding and is the only one of its kind in the world, together with a commitment to focus not just on lucrative projects.

M.M.: As for the legacy of the French school of molecular biology, it is still very much alive. I believe that we are still within the conceptual framework established by Jacob, Monod, Lwoff and their team during the molecular revolution. Regardless of the latest innovations, such as systems biology, the way we explain how organisms function remains the same.

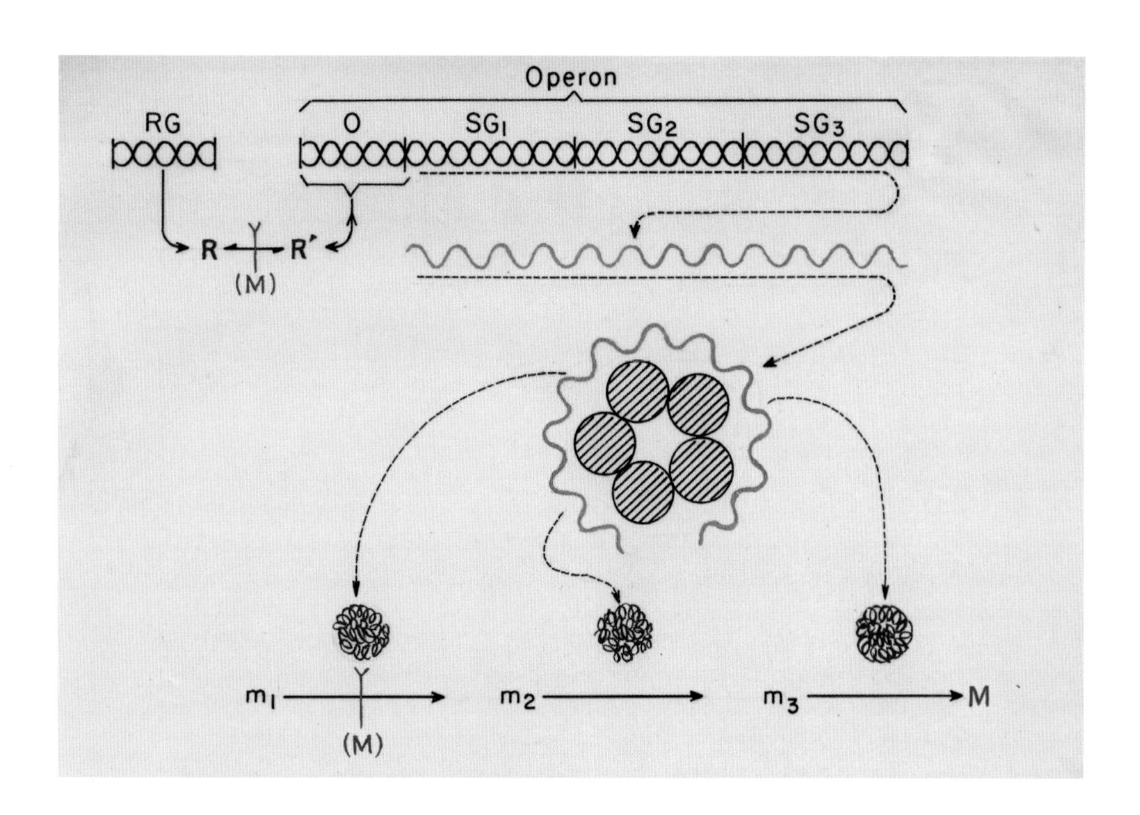

Lwoff, Jacob and Monod worked on the regulation of gene expression. Here we see the general structure of an operon, a DNA section that triggers gene transcription. (Archives of Jacques Monod, between 1961 and 1965)

The Institut Pasteur in the 21st Century

Would Louis Pasteur recognize his scientific legacy in today's Institut Pasteur? It seems a safe assumption to make, since that legacy has been carefully preserved over the years. Louis Pasteur's vision for his institute can be summed up in a single entreaty: he wanted his successors to transcend boundaries. He himself transcended the scientific boundaries between (bio)chemistry, microbiology and medicine, and his "disciples" soon followed in his footsteps, adding therapeutic chemistry and immunology. The Institut Pasteur is still a shining example of multidisciplinarity, and these more "traditional" disciplines have now been joined by new fields such as human genetics, neuroscience, and developmental and stem cell biology. At the intersection of these disciplines, dynamic new research areas have flourished. One such example is cellular microbiology, which combines viral and parasitic microbial genetics with cell biology. Pasteurians working in this field have developed a pioneering approach to decipher the virulence mechanisms used by pathogens. Louis Pasteur also transcended geographical boundaries, sending his disciples to the four corners of the world to establish centers of expertise for communities that were often disadvantaged and devastated by disease outbreaks. These centers soon became the regional antennae of an ambitious policy of outbreak surveillance and control. We still face the same challenges today, and the Institut Pasteur International Network, with institutes in 26 countries on every continent, remains on the front line. The International Network is also an invaluable asset for the Institut Pasteur, enabling it to tackle priority areas for health research at global level.

Drawing on its illustrious past, the Institut Pasteur continues to respond to current and future health challenges by investigating the physiological and pathological processes that characterize living beings and developing innovative diagnostic, therapeutic and prophylactic tools in the fields of vaccination and immunotherapy. The Institut Pasteur is constantly re-evaluating its priorities in a

Opposite: Photo taken in the Human Histopathology and Animal Models Unit directed by Fabrice Chrétien at the Institut Pasteur in Paris.

changing health environment characterized by the regular emergence of new infections, the growth of resistance to anti-infective drugs, environmental developments including global warming, its ecological consequences and challenges—which have major health repercussions—, and sociological phenomena such as globalization and an aging population, the latter spawning a host of consequences including a rise in cancer.

To tackle these challenges, the Institut Pasteur maintains high standards of scientific and technological excellence, ensuring it has all the tools it needs to unravel the complexity of biological systems. Technological progress is opening up new research fields, paving the way for innovative tools that can fight disease at the level of entire populations. This new approach, termed "global health", is employed in the ongoing crusade against emerging outbreaks. At individual level, these tools offer new treatment options and "precision medicine", a therapeutic concept tailored to each patient's genetic and physiological makeup and environment.

In view of these challenges, over the past few years the Institut Pasteur has significantly improved its capabilities in the "omics"—bioinformatics and biostatistics—and in integrative biology, which involves research at different levels (molecular, cellular and pathophysiological). It has also developed its expertise in disciplines that have become essential for biology, such as mathematics and computational biology (acquisition, analysis and storage of big data, statistics, modeling and artificial intelligence), chemistry (synthesis and analysis), physics (especially microfluidics and optics), and engineering (nanotechnology and synthetic biology). These new approaches offer fresh hope that we might one day be in a position to fully understand the incredible complexity of living organisms.

The main priority of the Institut Pasteur's research is investigating the fundamental properties of living organisms. This research is still based on microorganisms, which continue to serve as vital models to explore the basic molecular processes involved in cellular function. But over time it has extended to include a diverse range of eukaryotic model organisms, including Drosophila, zebrafish and mice.

Molecules, microorganisms and higher eukaryotes

Current research into the fundamental properties of living organisms, molecular systems and the mechanisms involved in normal and pathological conditions is partly based on reductionist principles. To determine the specific properties of microorganisms such as pathogens and commensals and explore how they interact with complex organisms such as humans, scientists are developing integrated, quantitative strategies, working upwards from molecules to cells, tissues, organs and the entire organism. This involves extensive research into the fundamental components of living organisms, both at molecular level (proteins, lipids and nucleic acids), using structural biology and molecular chemistry techniques, and at cellular level (prokaryotes and eukaryotes), with an agenda focused on cell biology.

Stem cells

Scientists are currently investigating the fundamental properties of stem cells—their plasticity and self-renewal mechanisms—with the aim of shedding light on the signals and cell interactions that regulate the balance between pluripotency and differentiation. By improving our understanding of the biology of these cells, we can provide answers to key questions about their physiology and their

1

Pluripotency: the ability of a cell to differentiate into all cell types.

microenvironment (niche) which still need to be elucidated before they can be used in regenerative medicine. The development of cellular models based on the generation of induced pluripotent stem cells (iPS cells) is opening up promising therapeutic and diagnostic possibilities, especially for autologous transplants and personalized therapeutic strategies. Given the increasingly aging population, cutting-edge academic research in stem cell biology is crucial, since age-related tissue deterioration is becoming a major clinical and societal problem. Once again, a thorough, detailed knowledge of the properties and behavior of stem cells, their niche and their modulation during aging and stress is vital for the development of new therapeutic strategies.

2

Autologous: when the donor and recipient are the same person (as opposed to heterologous transplants, which are between two different people).

The "Pasteur Tech Lab", based at the Institut Pasteur in Paris, hosts innovative technological R&D projects.

Epigenetics

Epigenetics refers to changes in gene activity and regulation that are not caused by variations in the DNA sequence. These changes, mainly controlled by alterations in chromatin conformation or the chemical modification of DNA and its associated proteins, occur during development, the immune response and infectious processes. Epigenetics is now a priority research field, with scientists investing considerable efforts in clarifying the links between epigenetic changes and phenotypic variability, in health and in disease. Genetic and epigenetic approaches are crucial for research into human variability in susceptibility to disease, and for improving our understanding of the co-evolution of human and microbial genomes.

Neuroscience and human health

Comprehensive approaches are also vital to help us understand how the human brain works and adapts to its internal and external environment. Normal brain function appears to depend on the close interaction between the immune system and the brain in homeostasis. Today's scientists use multidisciplinary methods to investigate the synergy between the brain and the immune system, exploring new theories about the pathophysiology of neurodegenerative and mental disorders that might lead to innovative therapeutic strategies. Human genetics can also help us understand the causes of neurological conditions such as autism and Alzheimer's disease. Close partnerships with clinicians in Parisian hospitals are especially valuable for research into these different branches of neuroscience.

How do hosts, microbes and vectors interact?

This question has always been central to the Institut Pasteur's work, but recent findings about the microbiota have taken research in this area to a whole new level.

Phenotypic: all the observable characteristics of an individual.

How microbial communities influence host health

In recent decades, scientists have demonstrated the fundamental role played by microbial communities such as biofilms and microbiota, which are found in all natural and human ecosystems and environments and can exert both harmful and beneficial effects. This realization has completely changed our view of the microbial world. Understanding how microbial communities shape their environment and influence the health of their hosts is a key challenge of modern microbiology. The two-way communication between the immune system and the microbiota affects several physiological pathways and can trigger pathological responses such as obesity, diabetes and cancer. Exploring the links between nutrition, the microbiota and the metabolism is also crucial for our understanding of how chronic diseases and cancer develop. New data have shown that there are functional interactions between the gut microbiota and the brain. In the past, scientists tended to investigate the interrelations between gut and brain to shed light on the regulation of digestive functions. But they are now realizing that the microbiota also has an impact on neuropsychiatric conditions such as mood disorders, as well as on social behavior.

Homeostasis: the biological process by which conditions (temperature, blood sugar level, etc.) are controlled to maintain levels that are beneficial for the body.

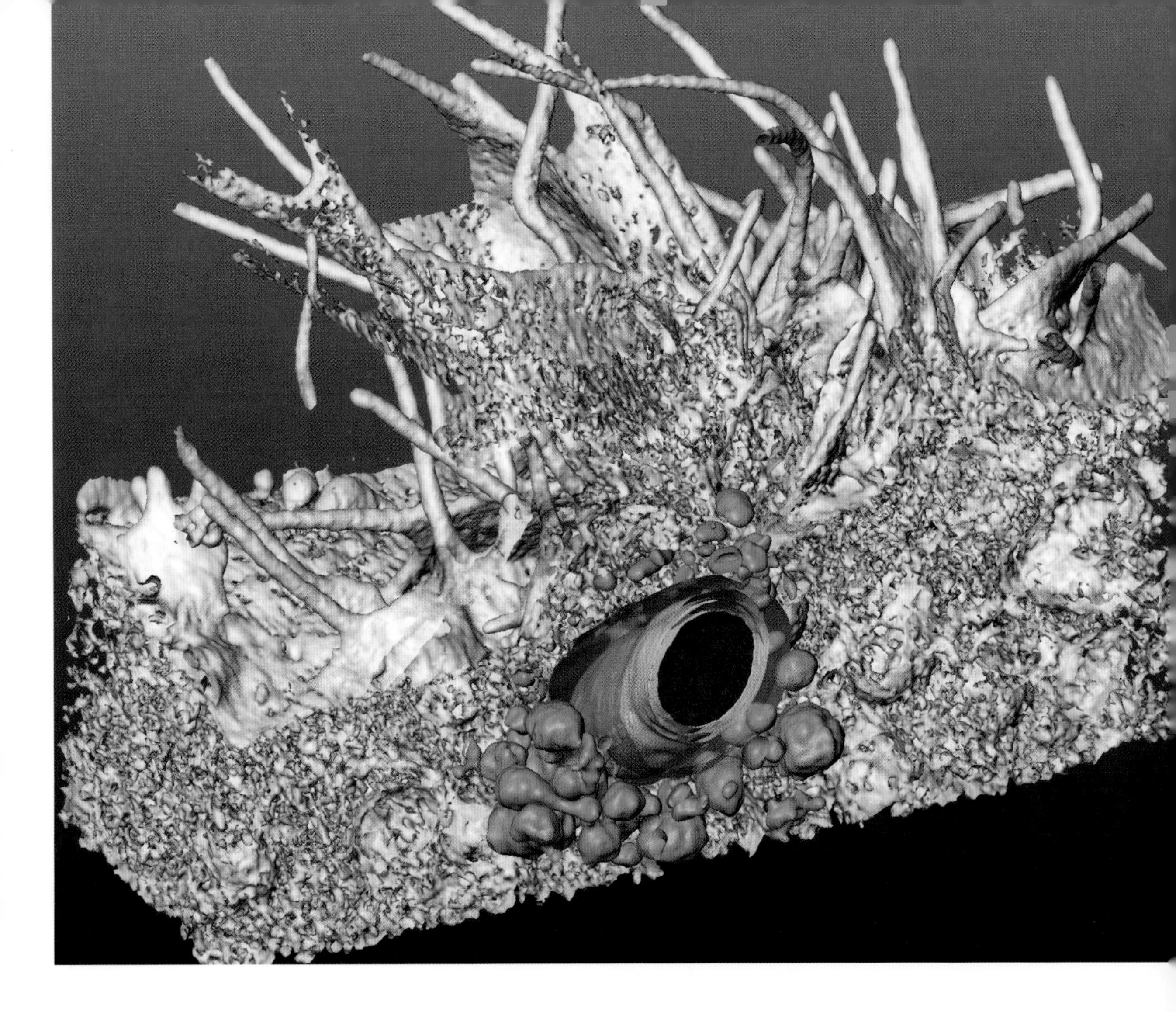

A number of new research fields have emerged to explore the role of the microbiota during homeostasis and infection and its resilience after infection, inflammation and antimicrobial therapy. Topics such as microbial assemblages, interbacterial communication and microbial communities are now the focus of considerable research. Another key research area is the virome, a term which refers to all the viral genomes in a single organism. Scientists believe that chronic viral infections play a decisive role in the immune response.

Cross-sectional image of a *Shigella flexneri* bacterium at the infection site in a HeLa cell. The bacterium is shown in blue, the macropinosomes in orange and the actin in gray. Correlative confocal imaging and focused ion beam scanning electron microscopy (FIB-SEM), reconstructed using Amira software.

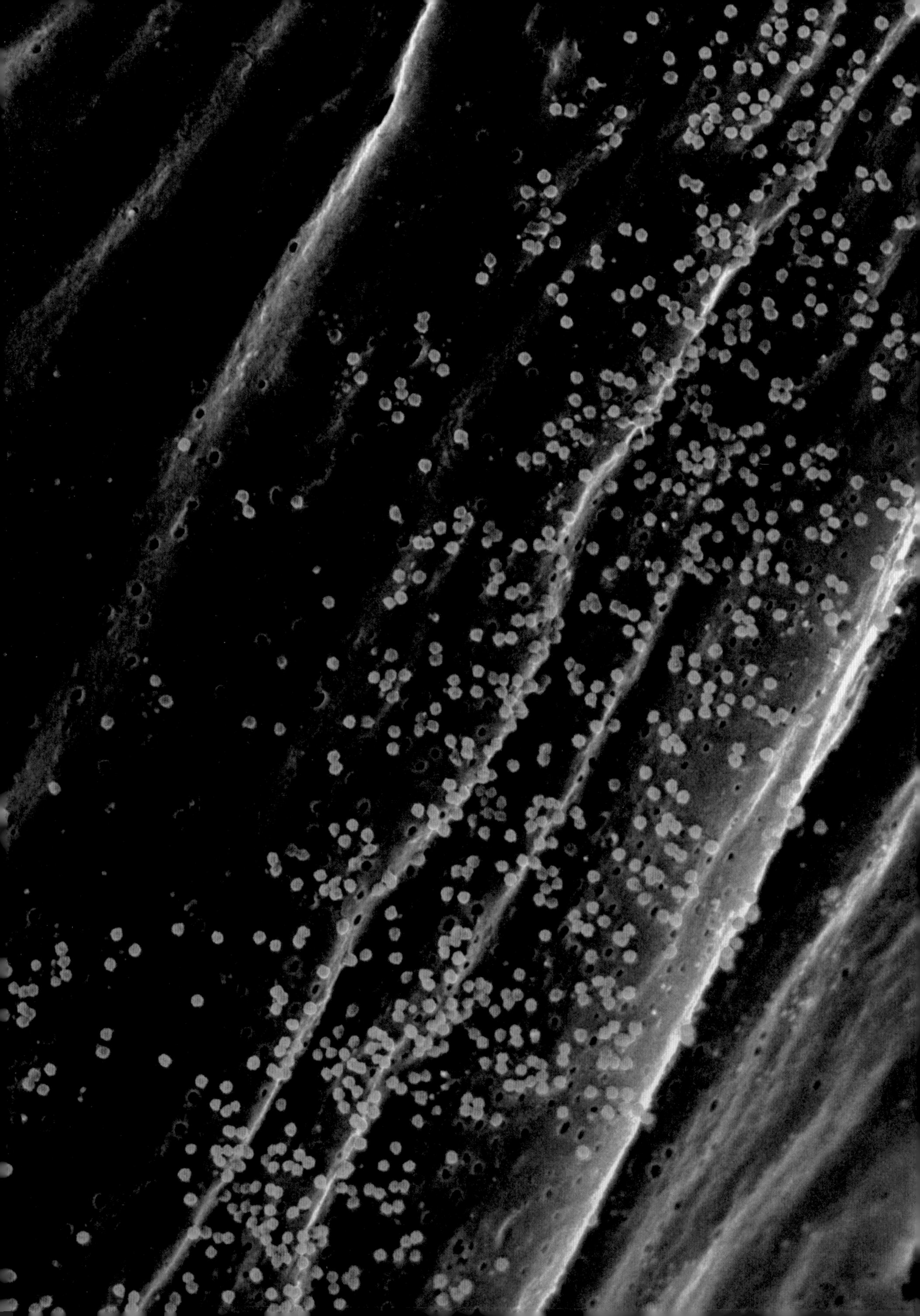

Devising novel antimicrobial strategies

Antimicrobial resistance is a huge challenge for global public health. The inexorable spread of resistant microbes represents a growing threat, and major efforts are still needed to help us understand the many factors that play a part in the emergence of resistance. Since this phenomenon is leading to a shortage in treatment options, scientists also need to focus their efforts on devising new antimicrobial therapies, especially antibiotics, and exploring or re-exploring concepts such as colonization by a carefully selected microbiota or phage therapy to target specific bacterial strains.

Research on arthropod vectors and the emergence of viruses

The past few decades have seen the emergence of new outbreaks of insect-borne viruses such as chikungunya and Zika, which have now been added to the traditional pantheon of pandemic-prone diseases such as malaria and dengue. Research on insects and other arthropod vectors is still a priority area. Several aspects are being explored: pathogen persistence in vectors, the immune reaction of vectors to pathogens, the host-vector relationship, and the impact of geographical origin on vector biology.

By reproducing the entire pathogen life cycle (in both insects and animal models) in a confined environment, we can examine how pathogens affect the immune system, taking into account the microbiota in both vector and host. This comprehensive approach can provide us with vital tools for anticipating and managing future disease outbreaks. Mathematical models based on experimental data can be used to characterize infectious processes in detail. This research also enables us to investigate how insecticides and environmental changes are affecting the transmission and spread of pathogens.

How can we deal with variability among individuals and populations?

The Institut Pasteur's scientists specializing in immunology, microbiology, cell biology, genetics and computational biology are pooling their efforts to gain new insights into the variability of the human immune response and pinpoint the causes and consequences of this variability. The ultimate goal is to develop precision medicine that is tailored to each individual's (epi)genetic, physiological and microbial makeup.

Variable immune responses among populations

One of the major challenges in preventing and treating infectious, inflammatory and autoimmune diseases is the extreme variability among individuals and populations. This variability can result in vastly different responses to therapies and vaccines from one individual to the next, with fluctuating levels of efficacy and different side effects. Immune disorders can also predispose some individuals to infection, inflammation or autoimmune conditions, or to the development of cancer. By investigating the cellular mechanisms involved in tumor development, scientists hope to shed light on the heterogeneity of cancer and prognostic variability.

Opposite: Scanning electron microscopy view of viral particles on the cell surface of Chikungunya virus-infected fibroblasts in culture. Colorized image: the cell outer membrane appears in gray and viral particles in orange.

The body's immune response also gives rise to major changes in metabolic processes, especially via immune mediators. Interactions between the metabolism and the immune system on several levels suggest that many of the complications associated with obesity are underpinned by pathogenic mechanisms. This raises hopes for new therapeutic approaches.

Given the complexity of the immune response, it has never previously been possible to identify the parameters that lead to a "healthy" immune system and its natural variability. Unraveling this enigma will involve developing predictive algorithms based on deep learning, a key strategy in this research field.

The quest for precision medicine

This research raises major new prospects for precision medicine: in the long term, scientists hope to be able to determine an individual's chances of developing a specific disease and predict the severity of a condition or the efficacy of a therapy, based not only on the genetics and lifestyle of an individual or population, but also on the presence or absence of key microbes. The Institut Pasteur is in a unique position to become a leader in this cross-cutting research topic that draws on microbiology, genetics, stem cell biology and immunology.

How can we predict and model human diseases?

The emergence of human diseases depends on interactions between human host factors, environmental factors and microbial factors. For biological processes such as aging, cancer and infections, it is currently virtually impossible to make an accurate phenotype prediction, given the huge number of molecules involved, the complexity of their interactions, their partly random nature, and the many uncertainties and different scales involved (nanometric to macroscopic) that prevent us from developing a quantitative description of all these aspects. A major challenge for scientists is overcoming this complexity so that they can predict a physiological or pathological phenotype from genotypes (of the individual and his or her microbiota), the environment and exposure to symbiotic or pathogenic microbes.

Phenotype: all the observable characteristics of an individual.

The Institut Pasteur is aiming to provide *in silico* predictions of the behavior of a cell system in response to pathological processes, based partly on the generation of genotype and phenotype data, and partly on the development of artificial intelligence methods tailored for use with biomedical data. High-throughput data generation techniques will be combined with approaches such as process miniaturization and automation, mathematics, modeling and artificial intelligence to produce phenotype data that can help predict how a cell or organ will respond to a pathological process. This approach relies on effective cooperation between the Institut Pasteur's biologists and health specialists and experts in mathematics, chemistry and physics. It also requires the use of high-throughput techniques such as sequencing,

***In silico*:** research performed using computer models.

mass spectrometry and robotic imaging, which provide ever increasing reams of data. Deep learning strategies will enable scientists to analyze this huge volume of data and develop predictive models.

LEADERS CHARTING THE COURSE FOR RESEARCH

INSTITUT PASTEUR PRESIDENTS:

1887 - 1895	Louis Pasteur
1895 - 1904	Émile Duclaux
1904 - 1933	Émile Roux
1934 - 1940	Louis Martin
1940 - 1941	Gaston Ramon
1941 - 1965	Jacques Tréfouël
1965 - 1966	Charles Gernez-Rieux
1966 - 1971	Pierre Mercier
1971 - 1976	Jacques Monod
1976 - 1982	François Gros
1982 - 1987	Raymond Dedonder
1988 - 1999	Maxime Schwartz
2000 - 2005	Philippe Kourilsky
2005 - 2013	Alice Dautry
2013 - 2017	Christian Bréchot

CHAIRMEN OF THE INSTITUT PASTEUR BOARD OF DIRECTORS:

1887 - 1892	Jean-Edmond Jurien de La Gravière
1892 - 1900	Joseph Bertrand
1900 - 1905	Henri Alexandre Wallon
1905 - 1907	Jacques-Joseph Grancher
1907 - 1917	Jean Gaston Darboux
1917 - 1933	René Vallery-Radot
1933 - 1940	Alfred Lacroix
1940 - 1966	Louis Pasteur Vallery-Radot
1966 - 1970	Georges Champetier
1970 - 1973	Claude Lasry
1973 - 1982	Pierre Royer
1982 - 1988	François Jacob
1988 - 1994	Marcel Boiteux
1994 - 1997	Bernard Esambert
1997 - 2003	Philippe Rouvillois
2003 - 2005	Michel Bon
2005 - 2011	François Ailleret
2011 - 2013	Jean-Pierre Jouyet
2013 - 2016	Rose-Marie Van Lerberghe
2016 - ...	Christian Vigouroux

Detail of a mass spectrometer, an instrument used in the Structural Mass Spectrometry and Proteomics Unit.

THE INSTITUT PASTEUR'S 11 SCIENTIFIC DEPARTMENTS

CELL BIOLOGY AND INFECTION

Scientists in the Department of Cell Biology and Infection focus their efforts on unraveling the intricate workings of microbes and cells. Their overriding aim is to shed light on the interactions between infectious agents and their hosts by examining the mechanisms governing normal and pathological cell function.

DEVELOPMENTAL AND STEM CELL BIOLOGY

The Department of Developmental and Stem Cell Biology covers a broad spectrum of research, ranging from individual cells to whole organisms and from embryos to adults. How do cells acquire their identity, and how are organs formed? The department's work in the field of developmental biology has given rise to research on stem cells and their potential role in tissue regeneration.

GENOMES AND GENETICS

Scientists in the Department of Genomes and Genetics use experimental and computer-based techniques to explore the nature of genetic information in organisms of increasing complexity, from bacteria and yeasts to humans. They also investigate the evolution of infectious microbes and the selective pressure they have exerted on human genes over time.

IMMUNOLOGY

Ever since the immune system was discovered, scientists at the Institut Pasteur have been captivated by its incredible complexity. They all share the same determination to explore fundamental immunological processes with the aim of tracing the origins of disease, inspiring the development of new vaccines and devising new therapeutic strategies.

INFECTION AND EPIDEMIOLOGY

The Department of Infection and Epidemiology studies all aspects of infectious diseases: reservoirs and transmission pathways of pathogens, virulence factors, pathophysiological processes in the host, the innate immune response and the role of vaccines. These research topics range from academic science to clinical research.

MICROBIOLOGY

Microorganisms are the most abundant and diverse life forms on the planet. Since bacteria are responsible for a vast number of infectious diseases, they have always been a research priority for the Institut Pasteur. They also serve as excellent models for exploring the fundamental aspects of cell life at molecular level and the mechanisms developed by their hosts to fight against or live with them.

MYCOLOGY

Over the last thirty years, fungal infections have become a major public health concern. The Department of Mycology focuses its research on the three main fungi responsible for invasive infections: *Aspergillus fumigatus*, *Candida albicans* and *Cryptococcus neoformans*. The aim is to shed light on the biology of these pathogenic fungi, identify their virulence mechanisms and develop new strategies for diagnosis, prevention and treatment.

NEUROSCIENCE

The Department of Neuroscience centers its research on the organization and workings of the central nervous system at all levels, from molecules to behavior. The department's scientists apply this fundamental knowledge to investigate pathological conditions including neurological diseases, behavioral disorders and sensory deficits, tackling major medical challenges ranging from hearing loss and autism to addiction, neurodegeneration and mood disorders.

PARASITES AND INSECT VECTORS

The department conducts research into three key eukaryotic parasites responsible for severe diseases of major health and economic burden in the world's most populous regions: *Plasmodium*, which causes malaria; *Leishmania*, the leishmaniasis agents; and *Trypanosoma*, responsible for sleeping sickness. The *Anopheles* mosquito, the vector of *Plasmodium* and various different viruses, is also studied, alongside the tsetse fly, the sleeping sickness vector.

STRUCTURAL BIOLOGY AND CHEMISTRY

The structure of a molecule is intricately linked to its function. The units in the Department of Structural Biology and Chemistry focus their research on the three-dimensional organization, properties and synthesis of molecules of biological interest, especially those that play a role in human disease. This research reveals vital information for the development of new therapeutic and vaccine strategies.

VIROLOGY

The Department of Virology focuses its research on viruses, investigating their molecular organization and pathogenicity determinants, their replication and their interaction with host defense mechanisms. Viruses under study include respiratory viruses such as influenza; viruses that cause cancer (papillomaviruses, HTLV, and the hepatitis B and C viruses); retroviruses such as HIV; insect-borne viruses responsible for severe diseases including dengue, chikungunya, yellow fever and Rift Valley fever; and viruses causing hemorrhagic fever (such as the Lassa fever and Ebola viruses).

An International Network
to Rise to the Challenge of Globalization

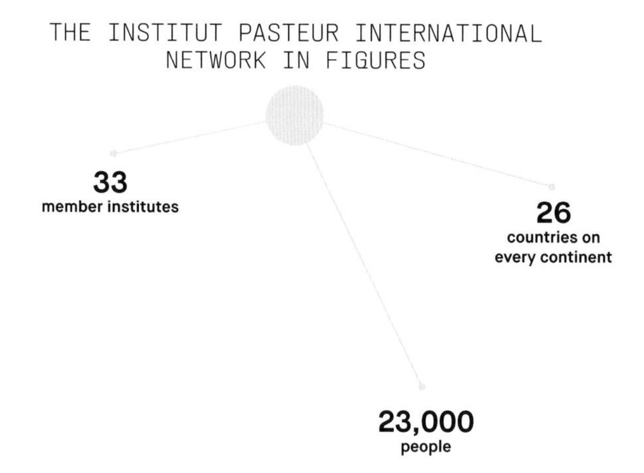

Louis Pasteur wanted to export progress in science to the colonial territories of the French Third Republic and beyond. He encouraged his "disciples" to travel the world on public health missions and fight epidemics in remote locations. The first International Network institutes outside of France derive as much from this determination as from the prestige surrounding the great scientist himself. But how did this man, who rarely left France, manage to foster an international network that, despite wars, decolonization, and economic and political crises, continues to develop today? For the answer, we need to look at the beginnings of this institution, which is quite unlike any other.

The Institut Pasteur in Laos was officially opened on Monday January 23, 2012, in Vientiane. It was the Institut Pasteur International Network's 32nd institute.

From East to West

The response to the rabies vaccine was so overwhelming that the most persuasive ambassadors of the Pasteurian cause were the patients treated in Paris. In December 1885, barely six months after the first trial of the vaccine, four children from Newark, who had been bitten by a rabid dog, were sent to Pasteur. Their return in perfect health was widely reported in the press and, as a result, two institutes were set up, one in New York and a short-lived one in Chicago.

Meanwhile, in the East, the czar sent Pasteur 19 Russians who had been attacked by a raging wolf. Three of them died but the other 16 returned home safe and sound. At the request of Alexander III of Russia, Pasteur sent his nephew, Adrien Loir, to set up an anti-rabies center in Saint Petersburg, the forerunner of the current institute in the city.

Adrien Loir was also sent to the Antipodes on another mission. In 1887, the government of an Australian province held a contest to curb out-of-control rabbit breeding and

The Institut Pasteur in Tunis, which opened in 1905. View of the front of the building.

twenty-five thousand pounds were offered in prize money. Pasteur came up with the idea of tackling rabbit numbers with fowl cholera bacteria, but by the time his nephew reached Sydney armed with cholera broth, the authorities had changed their mind. So Loir focused his efforts on a livestock illness, Cumberland disease, which he demonstrated was in fact anthrax. He opened an Institut Pasteur to produce an anthrax vaccine but production ended after his departure in 1893. There is no Institut Pasteur in Australia today but Loir's initiative meant 8 million sheep were inoculated without the Parisian parent institute receiving a single cent.

Overseas

Despite these false starts, the momentum was gathering and in the 20 years that followed the opening of the Institut Pasteur in Paris, many institutes were founded around the world. In 1891, Albert Calmette set up the Institut Pasteur in Ho Chi Minh-City, formerly Saigon. He adapted the smallpox vaccine preparation to young buffalo, introduced vaccination campaigns and studied antivenom serotherapy to treat cobra bites. Alexandre Yersin followed his example in Nha Trang in 1895, then other institutes opened up in Hue (1910), Hanoi (1921), Da Lat (1936), Phnom Penh (1953) and Vientiane (2012).

The first institute on the African continent opened its doors in Tunis in 1893 under the directorship of Adrien Loir, who had just returned from Australia. Charles Nicolle succeeded him in 1902. He turned the modest laboratory into a modern research center and highlighted the role lice play in the transmission of typhus—a discovery that earned him the Nobel Prize in Medicine in 1928. Institutes were then set up in Algiers in 1894, Tangier in 1910 and Casablanca in 1929. Émile Marchoux took Pasteurian methods to Sub-Saharan Africa. He founded an initial laboratory in Saint-Louis, the forerunner of the current

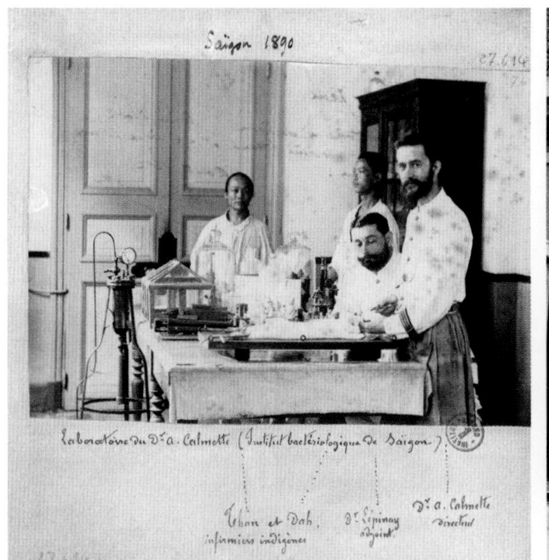

Above left: Albert Calmette and Dr. Lépinay in their laboratory in Saigon in 1891.

Above right: Entrance of the former Institut Pasteur in Cambodia on the Chroy Changvar peninsula in Phnom Penh, 1960.

Institut Pasteur in Dakar. But, besides these famous pioneers, hundreds of doctors, veterinarians, pharmacists and scientists were involved in the setup and success of institutes on every continent.

These spin-offs across the world came at a time when nations were fostering international cooperation in public health and this heralded the creation of the World Health Organization (WHO) in 1948. Ludwik Rajchman, a Polish-born Institut Pasteur doctor, was one of its main architects. Today, the Institut Pasteur International Network (IPIN) and WHO work closely together to monitor and fight infectious diseases.

Bound by an Ethics Charter

From the 1960s onwards, it became clear that the bilateral relationship between Paris and the institutes overseas needed to be reformed. Jacques Monod set up the Council of Directors of the Institut Pasteur and Associated Institutes in 1972 and, in doing so, laid the foundations for a network that would bring together independent facilities united by their shared goals and joint missions. Today, the IPIN has 33 members that have signed a cooperation agreement coordinated by the Institut Pasteur Department of International Affairs in

THE 33 MEMBER INSTITUTES
OF THE INTERNATIONAL NETWORK

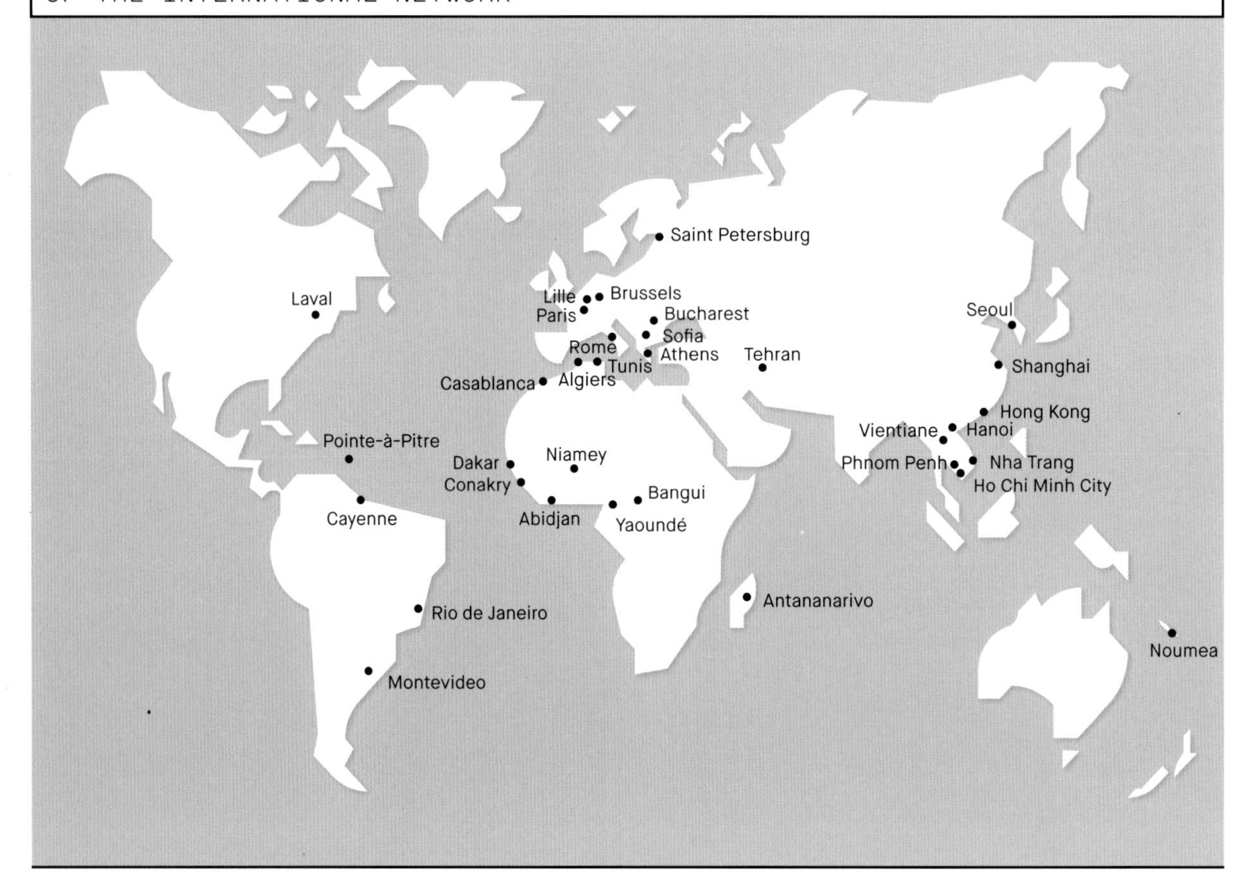

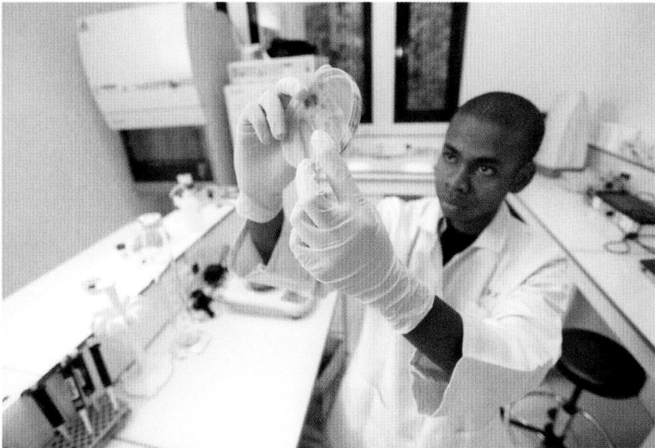

“To meet the challenges of world health, tools are needed for international research and global medicine. This is where the strength of the IPIN lies with 33 institutes in 26 countries, which are central to these challenge.”

Marc Jouan, Vice-President International Affairs and International Network

Paris. Its legal entity, Pasteur International Network, has two governing bodies—a General Meeting board bringing together all the institutes, and a Board of Directors, comprising 12 members. Like the institutes themselves, this governance structure is intended to be flexible and responsive to anticipate and adapt to health emergencies and the developments of the next 10 to 20 years. Although they differ in their facilities, statutes, organization and activities, the institutes are brought together by a charter of values advocating the cardinal Pasteurian virtues—integrity, devotion, rigor and critical thinking. This diversity is an asset as each institute is both a benchmark in its country, under the aegis of the Ministry of Health or Research, and a source of assistance, support and inspiration for the others.

The work of the institutes falls under four main areas—research, public health, services (medical testing, vaccination, clinics) and education. Over 60% of them are based in low- or middle-income countries where they fulfill their role in sometimes difficult conditions. The IPIN works hard to build their diagnostic, detection and emergency response capabilities. As they need to integrate all

Top left: Off-road vehicles from the Institut Pasteur in Bangui are used to bolster influenza surveillance efforts in areas outside the capital.

Top right: Laboratory in the Experimental Bacteriology Unit at the Institut Pasteur in Madagascar.

GLOBAL HEALTH CHALLENGES

The Center for Global Health (CGH), directed by Arnaud Fontanet, was officially opened in July 2014. It stems from the observation that health problems have become uniform worldwide, with a resurgence of infectious diseases in the Northern Hemisphere (AIDS) and an explosion of chronic diseases (diabetes, cardiovascular diseases, cancer, etc.) in the Southern Hemisphere. Similarly, the causes of disease—such as pollution or smoking—are the same. The fact that over 70% of emerging diseases are of animal origin (zoonoses) also justifies certain CGH initiatives. The One Health concept, which is based on a multidisciplinary approach, recognizes that human health also depends on animal health and the ecological balance.
The CGH has put together an international group of 16 experts to pinpoint exactly where the IPIN can make a valuable contribution to major global health challenges. One of the chosen themes is antibiotic resistance. Little is known about its geographical distribution in Asia and Africa and yet the most resistant strains are found in some of these countries, particularly India. As reference centers, around 15 International Network institutes receive numerous strains. These are studied and tested to determine their resistance levels. By cross referencing genotyping data and analyzing it using bioinformatics methods, unique information can be obtained about the spread, circulation and mechanisms of resistance around the world. Malaria, dengue and encephalitis have also been chosen by the experts and they bring together member institutes actively involved in these diseases.
In addition to the Outbreak Investigation Task Force, which specializes in epidemic emergencies (see page 92), the CGH runs a training program—the Pan-African Coalition for Training (PACT). Researchers from International Network institutes in Sub-Saharan Africa support local university lecturers and provide students with practical laboratory training.
This program also aims to foster exchanges by giving doctoral students and postdoctoral fellows the chance to complete internships in other IPIN institutes or other high-level international organizations in Africa.

The SEAe (South East Asia Encephalitis) team in Cambodia monitoring bat populations belonging to the *Pteropus* genus, the natural reservoir for the Nipah virus, in May 2015.

contemporary research (genomics, proteomics, bioinformatics, etc.) tools, technology transfer has been organized and training sessions have been introduced for managers, technicians and researchers throughout the world. To further strengthen links, every new researcher recruited in Paris must complete a three-month internship in one of the International Network institutes to get a taste of scientific, epidemiological and human realities in far-off lands.

Public health, monitoring and response to epidemics

Pathogens have no borders. In the 21st century, infectious diseases are a global threat and the IPIN plays an essential role in monitoring and detecting their emergence. Institutes in Vietnam and Cambodia, for example, were instrumental in shedding light on and fighting SARS (severe acute respiratory syndrome) and avian influenza. Following these health crises, the Lao government called for an institute to be set up in the country and the Institut Pasteur in Laos opened its doors in 2012. More recently, the institutes in Dakar, Côte d'Ivoire and Bangui joined forces to fight the Ebola outbreak. At the request of the Guinean authorities, the first stone of the future Institut Pasteur in Guinea was laid in Conakry in November 2016. As well as goals set out collectively, the IPIN responds to requests from local ministries and international organizations, particularly the WHO. These requests either go through the institutes of the given countries, or via intermediaries, such as the Outbreak Investigation Task Force, an international, cross-disciplinary unit dedicated to rapid outbreak response.

Forging ahead into the third millennium, the IPIN lives on thanks to the researchers and staff that it brings together on all the continents. They are all united by the same Pasteurian spirit and values.

Above: Search for mosquito larvae in French Guiana.
Below: The Institut Pasteur in Laos, June 2012. Laboratory work.

At the Forefront of Research

Tuberculosis,

a Scourge of the Past that Still Plagues Us Today

TUBERCULOSIS
IN 3 FIGURES

500,000
new cases of multidrug-resistant tuberculosis in 2015

1.8 MILLION
deaths each year

10 MILLION
new cases each year

Historians believe that in the 19th century, a third of all deaths in Europe were caused by tuberculosis (TB), the "white plague". The industrial revolution created the perfect conditions for the disease to reach epidemic proportions, with poor families living in cramped, overcrowded conditions in towns and cities. But it would be wrong to think that tuberculosis is a disease of the past. Despite major progress, the World Health Organization (WHO) estimates that in 2015 around a third of the global population were carriers of the tuberculosis bacillus, 10 million people went on to develop TB, and 1.8 million died from the disease, including 170,000 children. In people with HIV/AIDS, due to weakened immune systems, TB is the leading opportunistic infection, responsible for 35% of deaths. In the 1980s, this pathogenic partnership led to an upsurge in the number of cases.

Opposite: Plate showing six preparations of the tuberculosis bacillus.

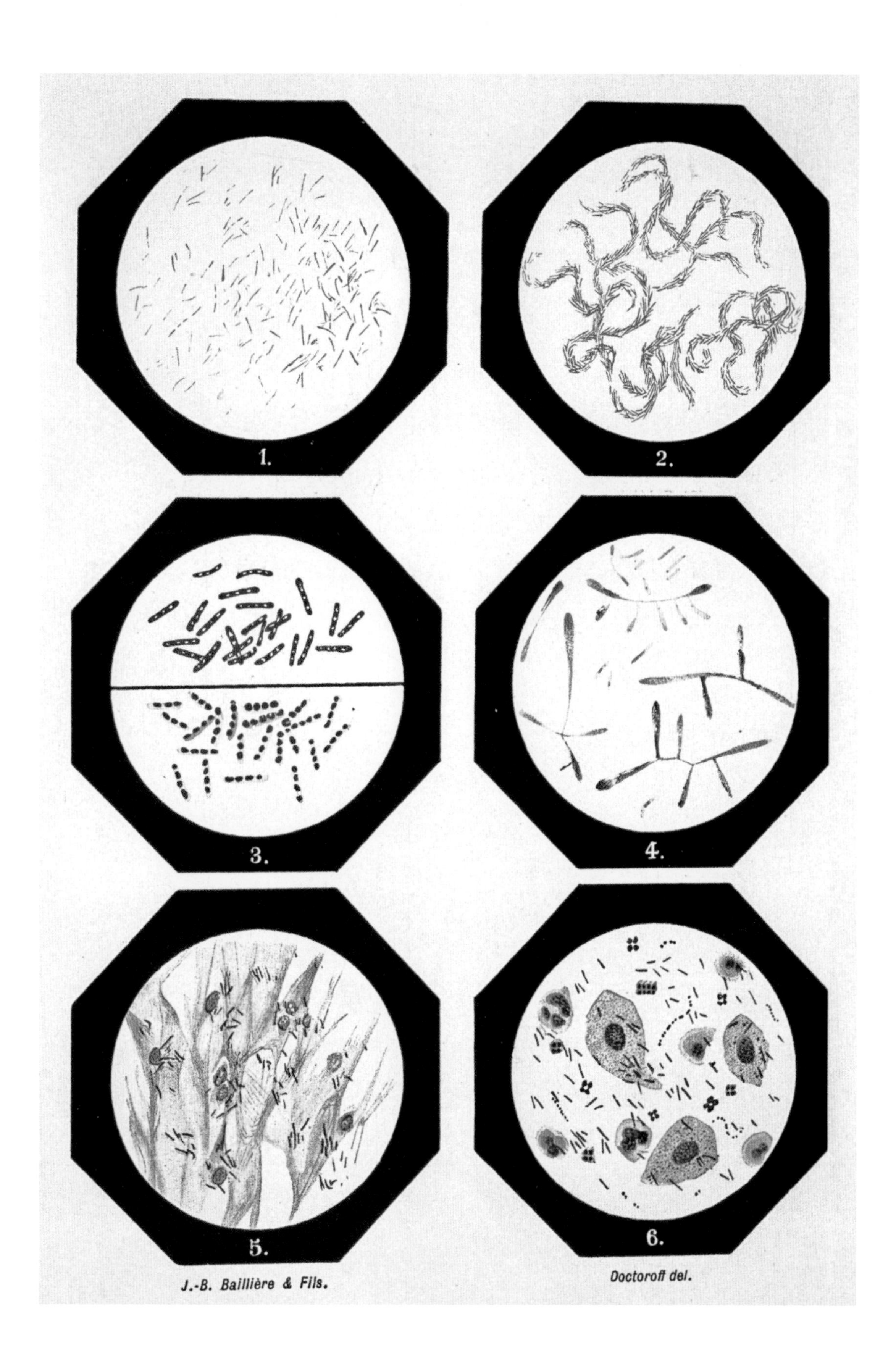
1.
2.
3.
4.
5.
6.
J.-B. Baillière & Fils.
Doctoroff del.

Poverty and poor living conditions have encouraged the spread of TB in low- and middle-income countries. Since the start of the 21st century, Europe has been the region where the disease has declined the fastest. In France, 4,741 cases were reported in 2015 (including 3,422 cases of pulmonary TB), confirming the slow but steady fall in numbers observed since 2008. Although WHO has set itself the target of eradicating TB by 2030, there is no cause for complacency, especially since a worrying threat has been looming on the horizon for nearly three decades now: multidrug-resistant TB (MDR-TB), in which bacteria are resistant to the two most effective first-line anti-TB drugs, isoniazid and rifampicin. Other strains that are highly resistant to two additional anti-TB drugs are known as "extensively drug-resistant" strains. In 2015, 2.8% of the reported cases of TB in France were caused by MDR-TB.

> "The Institut Pasteur has played a pioneering role in improving our understanding of infectious molecular processes with the development of cellular microbiology over the past 30 years."
>
> Philippe Sansonetti, Head of the Molecular Microbial Pathogenesis Unit at the Institut Pasteur

Stamps produced to support the fight against TB: "Science will win" (1953), "Early detection, quick recovery" (1949) and "The BCG offers protection against tuberculosis" (1948). These stamps are from the historical archives of the CNMR (French Committee Against Respiratory Diseases and Tuberculosis), which later became the Fondation du Souffle.

THE ANATOMY OF BACTERIA

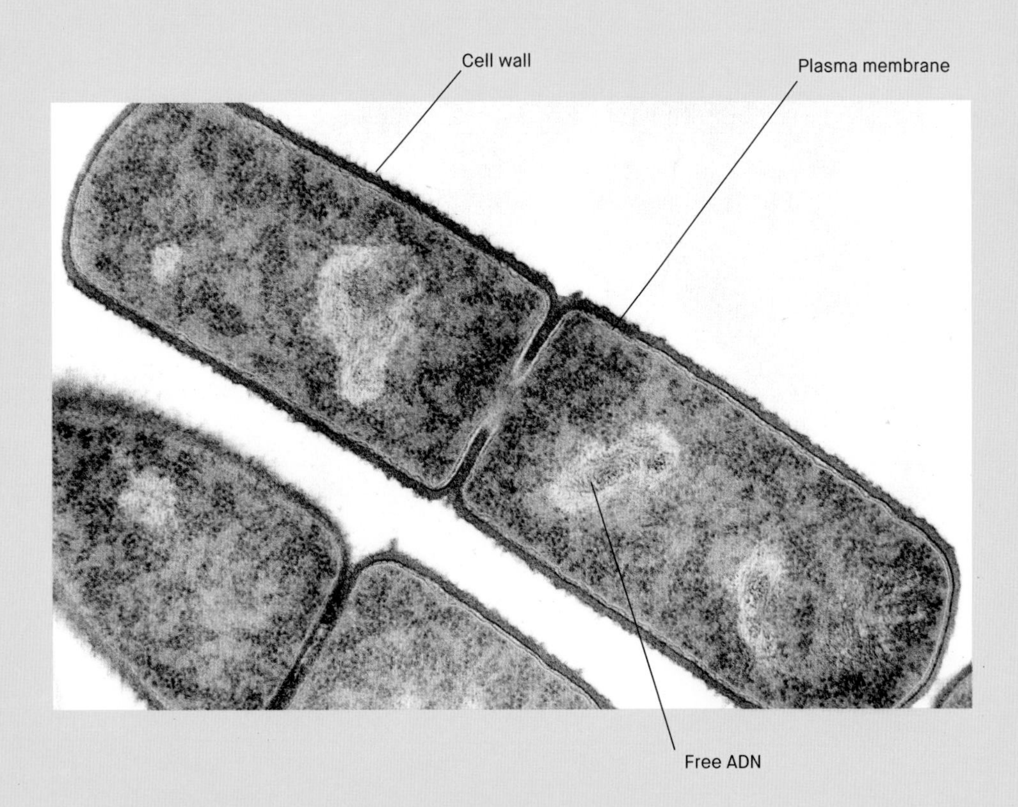

Bacteria were one of the objects of Louis Pasteur's research during his lifetime (1822-1895). They are defined as unicellular microorganisms with genetic material in the form of free DNA and may contain additional strands of DNA, especially circular molecules known as plasmids, which they are able to exchange with other bacteria. Most bacteria are protected by a wall that surrounds the plasma membrane. They reproduce by a process known as binary fission, in which a parent cell divides into two daughter cells.
Life on earth began with bacteria, and they are ubiquitous—they can be found in the ground, the ocean, the air we breathe, on and inside our bodies and even in the most hostile environments such as sulfur-rich hot springs and the ice of the Antarctic. They are invisible to the naked eye, ranging from 1 to 10 microns in size (approximately 1,000 times bigger than a virus). The vast majority of bacteria are not pathogenic for healthy individuals, but some can be fatal. Virulent bacteria have two main strategies for causing harm: they can either secrete toxic substances, like the bacteria that cause cholera and tetanus; or they can spread and invade one or more organs.
Scientists at the Institut Pasteur are currently using tools to get inside bacteria, examining their DNA to find virulence genes, tracing their history using phylogenetics, and analyzing the molecular dialog between pathogens and infected cells using microbiology techniques. Immunologists are focusing their efforts on investigating the defenses of the host organism.

Dividing *Bacillus subtilis* bacteria.

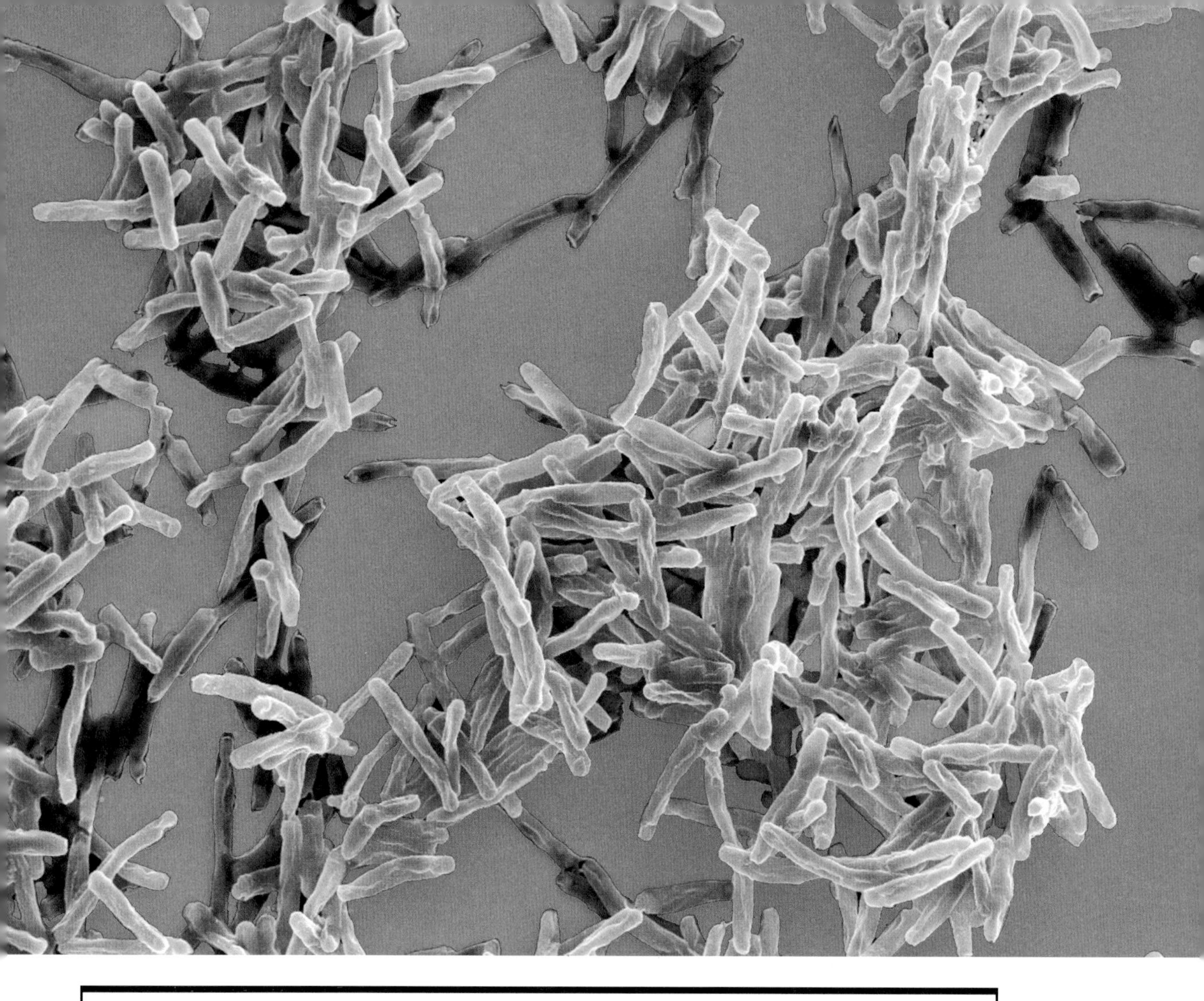

WHERE DOES *MYCOBACTERIUM TUBERCULOSIS* COME FROM?

M. *tuberculosis* is one of humankind's oldest companions; it is the result of a lengthy period of co-evolution with its host. Before genome sequencing, it was thought that *M. tuberculosis* was a variant of the tuberculosis agent in cattle, *M. bovis*, which had spread to humans during the Neolithic period, when animals began to be domesticated and farming populations came into closer contact with livestock. But the comparative genomics studies carried out by Roland Brosch and his team at the Institut Pasteur in Paris resolved this historical enigma by demonstrating that the animal strains are actually derived from a human bacillus closely related to *M. africanum*. *M. tuberculosis* is now believed to come from the same evolutionary branch as *M. canettii*, a rare species isolated in humans and confined to East Africa. The bacterium then acquired the virulence and resistance that enabled it to spread across the world, partly by losing the functions of some genes and partly by acquiring new genetic material transferred from bacteria belonging to a different species. Understanding the history of a given pathogen and the determinants of its adaptation to a host—in this case humans—can help us explain its pathogenicity and geographical distribution, and evaluate the risk of new outbreaks.

Mycobacterium tuberculosis, the tuberculosis agent (scanning electron microscopy).

A polymorphous disease

Tuberculosis is an infectious disease caused by a bacillus isolated in 1882 by Robert Koch, which was given the name *Mycobacterium tuberculosis* (*M. tuberculosis*), in line with nomenclature conventions. The disease can give rise to a variety of symptoms, but most often the bacteria actually go unnoticed, remaining in a dormant state in the host's body and often lurking in the fat cells of adipose tissue, as the Institut Pasteur's Mycobacterial Genetics Unit demonstrated in 2006. Only 5 to 10% of people infected with *M. tuberculosis* will actually go on to develop TB, either on first exposure to the bacterium (primary infection) or due to a reactivation of quiescent TB. The most common form is pulmonary TB, which has gone by many different names over the centuries, from the "phthisis" described by Hippocrates to the "consumption" of the 19th-century romantics. But TB bacteria can also infect the skin, bones (Pott's disease in the vertebrae), urinary tract, central nervous system (meningitis), digestive tract and lymph nodes, the famous "scrofula" that the kings of England and France were believed to cure with the "royal touch".

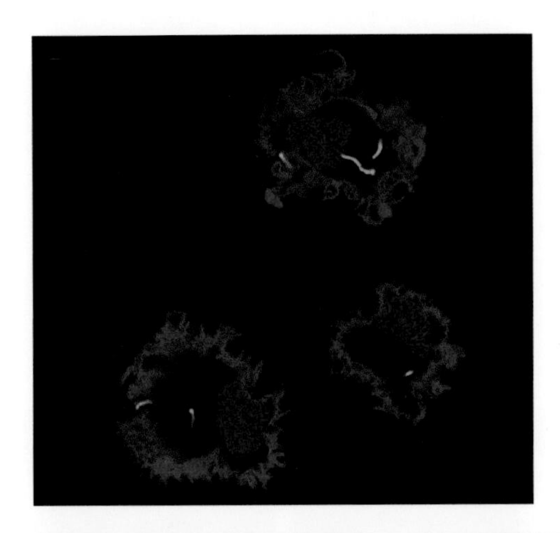

A MOLECULAR DIAGNOSTIC TECHNIQUE WITH BENEFITS FOR EPIDEMIOLOGY

Confirming a tuberculosis diagnosis involves isolating the bacterium in bronchial secretions via a sputum test. If *Mycobacterium tuberculosis* cannot be directly identified with a microscope it takes 2 to 4 weeks to obtain the results of a culture test, given that *M. tuberculosis* multiplies so slowly. It was at the Institut Pasteur that scientists first demonstrated that the bacterium could be identified more quickly by extracting DNA, amplifying regions of interest using PCR (polymerase chain reaction) and performing hybridization. They then targeted a genome region made up of repetitive mobile insertion sequences known as IS6110. Since each strain is characterized by a number of repetitions and their position, this technique can also be used by epidemiologists to monitor the spread of a specific bacterial line.

The tuberculosis bacillus (in green) entering dendritic cells (in red). Confocal microscopy, as used here, has played a major role in some recent discoveries about the disease.

TB is an airborne infectious disease, spread by droplets of saliva containing the bacillus that are released into the atmosphere when infected people cough or sneeze. The bacteria always attack the lungs first. From the lungs, they sometimes then spread to other organs, where they can develop. People with latent TB are not infectious, but those with non-treated active pulmonary TB will infect on average 10 to 15 other people each year. People in a weakened state or suffering from malnutrition, and especially those with compromised immune systems such as people with AIDS, are more at risk of developing TB when they come into contact with the bacillus.

Strategies adopted by insidious bacteria

Mycobacterium tuberculosis belongs to the *Mycobacterium* genus, like the leprosy agent, *M. leprae*. Other bacteria share more than 99.9% of their genomic DNA with *M. tuberculosis*, including *M. africanum*, endemic in West Africa, and *M. bovis*, the pathogen that causes bovine TB and can also infect humans. Together with other specific animal strains, these pathogens are referred to collectively as the "*M. tuberculosis* complex". The *M. tuberculosis* genome, sequenced in 1998 by Stewart T. Cole's team at the Institut Pasteur, contains approximately 4,000 genes. The chromosome develops freely in a cytoplasm surrounded by a complex lipid-rich envelope, which offers effective protection against the antibacterial activity of infected cells. In tuberculous lesions, *M. tuberculosis* is found both inside and outside cells of the affected organ; it is therefore known as a facultative intracellular parasite.

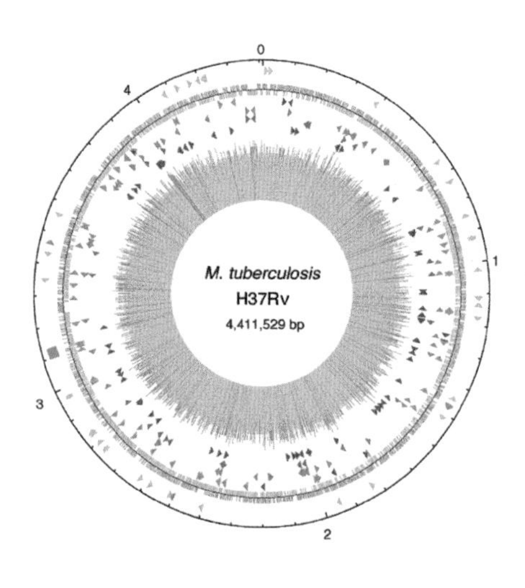

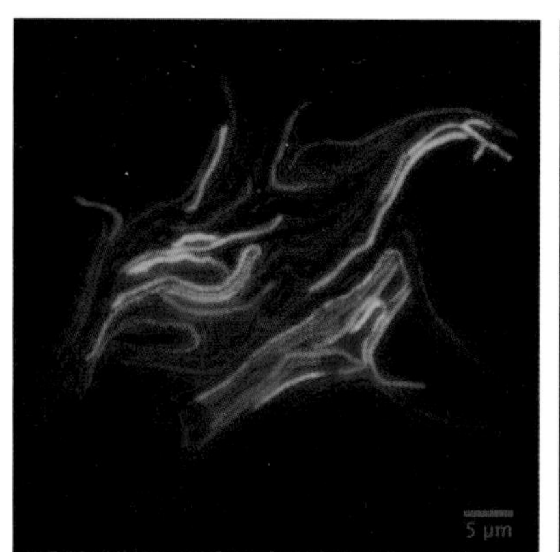

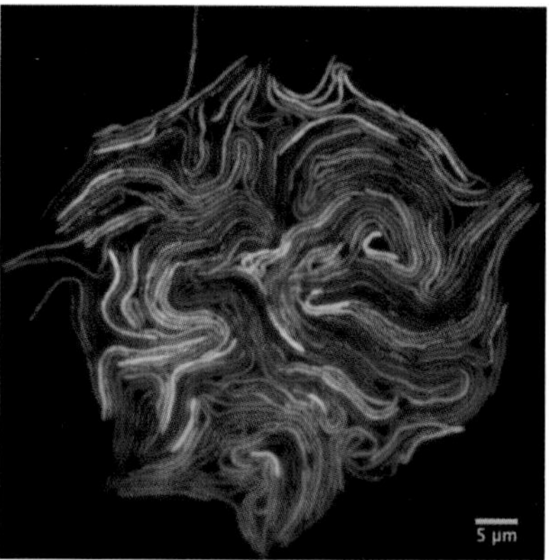

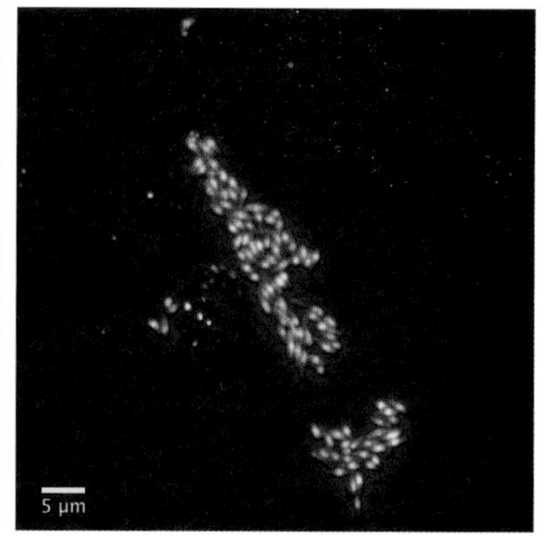

Top right: Circular map of the *Mycobacterium tuberculosis* chromosome, which was sequenced in 1998 by the teams led by Stewart T. Cole at the Institut Pasteur and Bart Barrell at the Wellcome Trust, UK.

Above left and center: Images of *Mycobacterium smegmatis*, a non-pathogenic model for tuberculosis, viewed in fluorescence. When exposed to an antibiotic that inhibits DNA replication, the bacteria lengthen.

Above right: Images of *Mycobacterium smegmatis*. Yellow fluorescence is a marker of DNA replication.

This strictly human bacterium has two characteristics rarely seen together, which have enabled it to travel the globe: it is spread through the air (via infected droplets from pulmonary lesions in infected people), and it can remain dormant in the host. It acts by interfering with the host's immune system. *M. tuberculosis* infects macrophages, cells in the body's first-line defense mechanism that ingest and destroy foreign particles via a process known as phagocytosis.

1

Macrophages: large cells in the immune system. As the body's first line of defense, they are responsible for ingesting and destroying viruses, bacteria, damaged or old cells, etc.

The bacillus has developed molecular strategies to persist and multiply within macrophages, which it uses to spread throughout the body. Two avenues of research are helping shed light on these strategies: the first focuses on identifying virulence genes, and the second examines interactions with the host. Drawing on the first whole genome analyses carried out at the Institut Pasteur in the 1990s, the team in the Mycobacterial Genetics Unit, directed by Brigitte Gicquel, identified a chromosome region that triggers the production of proteins known as dimycocerosates, which are involved in bacterial virulence. The scientists demonstrated that these molecules reduce macrophage production of two cytokines that modulate immune cells, TNF-alpha and interleukin 6 (IL-6). This enables *M. tuberculosis* to circumvent the body's early response to infection.

Cytokines: molecules produced by macrophages and other cells in the innate immune system that are involved in infection, inflammation, immunity and cell reproduction.

NETWORKING TO COMBAT TUBERCULOSIS

In 2000, the TB Vaccine Cluster consortium, comprising specialists in a variety of disciplines including chemistry, biology, immunology and genetics, was set up by Brigitte Gicquel. The network gave rise to several projects, including TBVI (the Tuberculosis Vaccine Initiative), a foundation with a consortium of around 40 partners, mainly in Europe. Several vaccine candidates currently undergoing testing have been developed as a result of these initiatives, which proved an effective means of coordinating TB vaccine research at European level.
OFID (the OPEC Fund for International Development) is an international network funded by the Organization of the Petroleum Exporting Countries. It facilitated technology transfer to leading laboratories in a dozen countries in Central and East Africa, including three Institut Pasteur International Network institutes. As well as using this new technology to improve their diagnostic tools, some of the laboratories have been involved in clinical trials for the development of short-course treatments for multidrug-resistant TB.
Finally, adopting an approach already developed for a number of targeted cancer therapies, the NAREB network (Nanotherapeutics for Antibiotic Resistant Emerging Bacterial Pathogens) coordinated by Brigitte Gicquel aims to use nanoparticles to transport TB drugs to infection sites so as to improve their efficacy and reduce unwanted side effects.

Using transcriptomics, Ludovic Tailleux and his colleagues in the Host Response to Bacterial Infection Group revealed another strategy employed by the bacterium. They isolated 33 genes whose expression profile is strengthened in infected macrophages. These genes, which contribute to the formation of new blood vessels, especially by producing VEGF (vascular endothelial growth factor), are used by the bacterium to spread throughout the body from the infection site. VEGF inhibitors have been developed to treat various types of cancer, and these could also be tested for tuberculosis. More recently, the Microbial Individuality and Infection five-year group (a research team set up for a five-year period and directed by promising young scientists), led by Giulia Manina, developed a new exploratory approach. The team is using innovative tools such as real-time microfluidics and epifluorescent microscopy to examine the bacilli in a colony individually in a bid to understand how the tiny variations observed between them influence quiescence, persistence and renewed virulence.

Transcriptomics: a technique that examines messenger RNA, the temporary copies of sections of DNA that are produced during genome transcription and used for protein synthesis.

Microfluidics: a technique used to manipulate fluids at micrometer scale. In biology, it enables scientists to perform experiments on single cells and to process large numbers of samples.

Epifluorescent microscopy: a technique used to observe fluorescence that is not visible with an optical microscope. This fluorescence is naturally emitted, either by biological substances in cells, or by structures that have been marked with fluorescent dyes.

> "Multidrug-resistant tuberculosis is more than a threat; it is a glaring problem which, in this age of international travel, concerns every country on the planet."
>
> Brigitte Gicquel, Head of the Mycobacterial Genetics Unit at the Institut Pasteur

NATIONAL REFERENCE CENTERS AT THE FOREFRONT OF COMMUNICABLE DISEASE SURVEILLANCE

National Reference Centers (CNRs, see page 176) serve as both expert microbiology laboratories and observatories for monitoring communicable diseases. At national level, they centralize information and contribute to efforts to fight these diseases. The Institut Pasteur hosts 14 of France's 44 CNRs, including the Tuberculosis CNR. They play a key surveillance role in the fight against infectious diseases. Their mission can be divided into five main strands: microbiological expertise and susceptibility surveillance; epidemiological monitoring; keeping public authorities informed of any unusual phenomena, from clusters of cases to rising numbers; providing advice to the French Health Ministry, Santé Publique France and health agencies; and offering training through knowledge transfer, hosting interns, running conferences, etc.

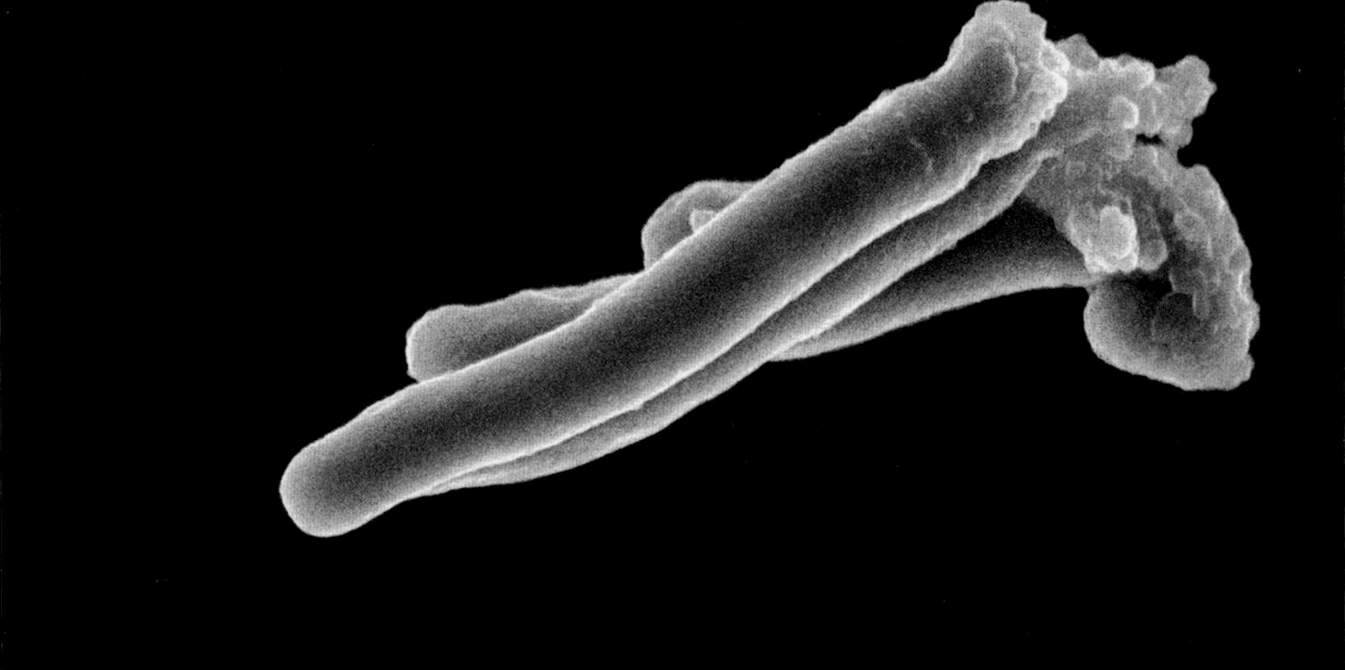

Overcoming resistance

Tuberculosis is treated with a combination of antibiotics prescribed for at least six months. Incomplete or erratic treatment leads to the emergence of resistant strains, which can then spread among the community. These resistant cases previously required longer and far more costly therapeutic strategies, but in 2016 WHO approved the use of a less expensive, shorter regimen (9 months). Given the rise in bacterial resistance, the development of innovative anti-TB drugs is now an urgent public health priority.

The Institut Pasteur in Paris and the Institut Pasteur of Shanghai are using a combination of genetic and genomic approaches and high-throughput screening to identify new molecules that can play an active role in multidrug-resistant tuberculosis. In Lille, teams from the Institut Pasteur, the CNRS and Inserm have updated a traditional tuberculosis drug, ethionamide. This antibiotic only becomes active when it penetrates the bacterium, after being activated by one of its enzymes. Resistant forms of TB have developed a mutation that blocks the action of this enzyme. To overcome this problem, the scientists have designed a small molecule, SMARt-420, which replaces this defective protein. The combination of ethionamide and SMARt-420 has been found to be effective in mice; human trials are due to begin shortly.

The tuberculosis bacillus (*Mycobacterium tuberculosis*) is thin, rectangular and immobile. In culture it appears as rods.

EARLY TRIUMPHS FOR CHEMISTRY OVER INFECTIOUS DISEASES

In 1935, German pathologist Gerhard Domagk demonstrated that a synthetic product patented as Prontosil could be used to treat streptococcal septicemia induced experimentally in mice. A few months later, Jacques and Thérèse Tréfouël, Federico Nitti and Daniel Bovet, from the Therapeutic Chemistry Department at the Institut Pasteur, discovered that once the product was inside these animal models it actually split into two molecules: one colored and inactive, the other colorless and effective in treating infection. The efficacy of this sulfonamide was confirmed when a child with streptococcal meningitis was successfully treated at the Institut Pasteur hospital. Chemical processing was soon used to create thousands of other similar compounds from this molecule —it is thought that more than 5,000 derivatives were available worldwide in 1945.

Sulfonamides were tested for their efficacy in treating tuberculosis, but the results were inconclusive. However, further research on this family of anti-infective compounds led to the development of isoniazid, the first recognized tuberculosis drug, which helped stem the tuberculosis epidemic that was still widespread in Europe after the Second World War.

Anti-infective chemotherapy using sulfonamides also proved effective in other conditions, including the skin infection erysipelas and the puerperal fever it can cause following childbirth. This progress also led to tests to assess the susceptibility of the pathogen, the forerunners of today's antibiograms. Sulfonamides gradually fell out of favor when penicillin, discovered by Alexander Fleming in 1928, proved to be effective in treating so many infectious diseases.

Culture of *Mycobacterium tuberculosis*, the human tuberculosis agent.

THE INSTITUT PASTEUR AT THE CUTTING EDGE OF TB VACCINOLOGY

Before the first effective chemotherapy approaches were discovered for TB, the search for a vaccine was a top health priority. When Albert Calmette returned from Indochina to France, he headed for the new Institut Pasteur in Lille and began his research by emulating Louis Pasteur's approach: his aim was to produce a weakened, stable bacterial strain. Two years later he was joined by veterinarian Camille Guérin. Calmette successfully cultured *Mycobacterium bovis*, the bovine TB pathogen, on potatoes soaked in glycerinated ox bile. Thirteen years later, after transferring the strains 230 times between successive culture media, the two men managed to obtain a highly stable attenuated strain, which they named Bacillus Calmette-Guérin (BCG).

The vaccine was first administered on July 1, 1921, to an infant whose mother had tuberculosis. But its use initially remained relatively limited, and it suffered a major setback in 1930 with the Lübeck tragedy, in which several children died after being given vaccines supplied by the Institut Pasteur. Following a high-profile trial, the German laboratory was found guilty of contaminating the batches with a virulent bacillus. In 1950, the BCG vaccination became compulsory in France for all infants and adolescents. It was removed from the list of compulsory vaccinations in 2007 but is still recommended for some at-risk populations and for inhabitants of the Greater Paris region and French Guiana. These new guidelines reflect the limited efficacy of the vaccination in adults and the decline of TB in France.

TOWARDS A NEW GENERATION OF VACCINES

But given the ongoing pandemic and the emergence of multidrug resistance, the Mycobacterial Genetics Unit is focusing its efforts on designing a new generation of vaccines. While Louis Pasteur neutralized the danger of pathogens using microbial culture techniques, today's scientists are developing bacteria that are lacking certain virulence genes. Laleh Majlessi and her colleagues in the Integrated Mycobacterial Pathogenomics Unit, directed by Roland Brosch, performed a detailed examination of the differences between the *Mycobacterium tuberculosis* genome and the BCG genome sequenced in 2003 by Stewart T. Cole's team. These bacteria have a very dense cellular envelope and use transport systems known as ESX to enable proteins to cross the highly selective barrier. Without these systems, the bacillus is no longer virulent, as in the case of the BCG, which has lost a functional ESX-1. Working in collaboration with the University of Pisa, the scientists developed a strain of *M. tuberculosis* lacking five proteins transported via the ESX-5 system, in which the transporter itself remained active and performed its function for other molecules, especially the PE/PPE proteins recognized by the immune system. Compared with the BCG, this vaccine was more effective in controlling pulmonary lesions in animals. Clinical trials on humans are due to begin shortly.

Other scientists have been investigating the phoP gene, which controls hundreds of other genes, some of which code for virulence factors. Researchers at the Institut Pasteur have produced a strain that has been stripped of its pathogenicity due to a mutation in the phoP gene. Preclinical trials led by the University of Zaragoza have revealed that this strain offers more effective protection than the BCG, with fewer unwanted side effects. A phase II clinical trial is currently under way in South Africa. Six European and South American laboratories coordinated by teams led by Brigitte Gicquel at the Institut Pasteur in Paris and Carlos Martin at the University of Zaragoza were involved in this research as part of the integrated TB-VAC European project.

Other research on antibacterial vaccines is currently ongoing. The Molecular Microbial Pathogenesis Unit directed by Philippe Sansonetti is particularly focusing its attention on the molecular and cellular bases of pathogenicity in *Shigella dysenteriae*, a bacterium responsible for potentially severe diarrhea. The scientists compiled a collection of strains with reduced virulence, administered orally as a pentavalent vaccine that is currently being tested in phase II clinical trials. The Chemistry of Biomolecules Unit directed by Laurence Mulard adopted another approach by developing subunit vaccines based on antigens known as bacterial lipopolysaccharides, which are targeted by antibodies. Clinical trials have begun in Israel.

VACCINATION, A MAJOR ISSUE

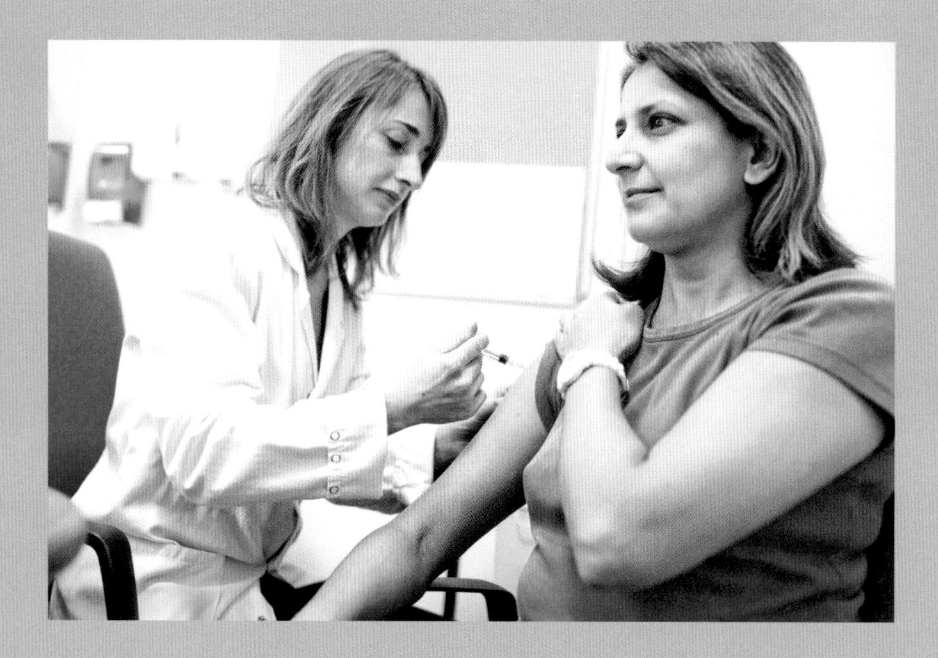

Vaccinology has a long history at the Institut Pasteur and remains a top strategic priority. More than a century after Louis Pasteur's death, infectious diseases are still among the leading causes of mortality worldwide. In particular, diarrheal diseases, tuberculosis, malaria, and HIV rank among the top 12 causes. Growing antimicrobial resistance is making the search for new vaccines all the more urgent. Addressing this need, several Institut Pasteur teams are actively developing novel vaccine technologies and approaches. Non-infectious diseases, such as cancer, follow closely behind in the ranking of leading causes of death and are also tackled with vaccines.

HOW DO VACCINES WORK?

Vaccination involves injecting an inactivated infectious agent or a fragment of that agent into a healthy person, without triggering the disease but eliciting an immune response that the body will remember. Macrophages, which are part of the body's first-line defense system—innate immunity—are capable of ingesting the invader and presenting its characteristic molecules, known as antigens, to other cells in the immune system. This activates B lymphocytes, which differentiate into plasma cells. These in turn produce antibodies that target the infectious agent. When some T lymphocytes directly recognize or are presented with antigens, they become cytotoxic and develop the ability to eliminate target cells. Others remember the infection and enable the body to activate its defenses more swiftly and effectively in the event of a future invasion. The principle of vaccination is based on this notion of cell memory, which sometimes needs to be reactivated by a booster injection.

A vaccination consultation at the Institut Pasteur Medical Center.

INFECTIOUS DISEASES

At the Institut Pasteur, the major focus of vaccinology research is on infectious diseases. Long-standing efforts in vaccine development have targeted shigellosis, a potentially serious diarrheal disease with a high mortality rate in children in low- and middle-income countries. The current vaccine candidate was developed by the teams of Laurence Mulard, Armelle Phalipon and Philippe Sansonetti and is under clinical evaluation with results expected for the end of 2017. Using the innovative synthetic approach to identify the best antigen, the vaccine was designed to mimic the bacterium's complex lipopolysaccharide, which is the main target of protective immune responses.
Major transversal research programs have been established in the fields of malaria (see page 63) and tuberculosis (see page 46), encompassing epidemiology, academic research, and vaccine approaches. For malaria, vaccine research includes the identification of novel antigens (Rogerio Amino), investigations to enhance the generally low immunogenicity of malaria antigens as vaccines (Chetan Chitnis), and novel approaches with live attenuated parasites (Salaheddine Mécheri). For TB, Laleh Majlessi, Roland Brosch and their teams have developed candidates that are currently under pre-clinical evaluation and could potentially replace BCG as more effective and longer lasting vaccines.
Another disease of the past, plague—or the Black Death—has attracted attention in terms of vaccine development due to its almost-global re-emergence and the fear of bioterrorism it raises. The vaccine candidate developed by the team of Christiane Demeure and Élisabeth Carniel is based on a close relative of the plague bacterium that is naturally far less virulent. The candidate vaccine fully protects mice against both bubonic plague and the highly lethal pneumonic plague, which supports the case for further development of the vaccine.

Existing vaccines are also the focus of research at the Institut Pasteur. For example, the first-ever vaccine for Dengue fever is only approved for ages 9 and above. As it is ineffective in younger children, teams at the Institut Pasteur are searching for more effective antigens. Anavaj Sakuntabhai and his team have identified new protective mechanisms by studying individuals who are infected but asymptomatic. And Félix Rey's team used structural analysis to identify specific viral surface regions that are needed to elicit protective antibody responses.

EMERGING DISEASES

Another key priority area is the development of vaccines against emerging diseases, as seen with the Ebola and Zika outbreaks. Novel technologies are needed to ensure rapid development of vaccines for these new diseases. Following in the tradition of Louis Pasteur who devised the first general method to develop vaccines against various pathogens, several teams are active in this field. One line of research concerns viral vectors that are non-virulent variants of viruses to which genetic material from a specific pathogen is added. This allows the vectors to trigger immune responses against that pathogen and makes them highly versatile platforms. Frédéric Tangy and his colleagues developed a new technology based on the measles vaccine, driven by the excellent safety record and efficacy of this childhood vaccine. The respective candidates against Chikungunya and Zika virus, licensed to Themis Bioscience, are among the most advanced for these diseases. Pierre Charneau's team has developed a technology based on a lentiviral vector which is derived from HIV but cannot spread. Designed initially for therapeutic use as the HIV vaccine licensed to Theravectys, the team has developed a modified vector with enhanced safety characteristics for use as a preventive vaccine. Additional viral (Marco Vignuzzi, Sylvain Baize) and bacterial (Gérard Eberl) technologies are in the pre-clinical development stage.

CANCER

A therapeutic breast cancer vaccine candidate, developed by the teams of Claude Leclerc and Sylvie Bay, is one of the Institut Pasteur's most promising vaccine candidates, based on a novel synthetic approach. Short chains of sugar molecules that are identical to the carbohydrates on cancer cells are synthesized and coupled to a protein carrier to enhance their potential for developing antibodies against the carbohydrates, and hence the cancer cells. A clinical trial is currently ongoing.

VACCINATION IN THE INTERNATIONAL NETWORK

There are many more vaccine projects ongoing within the Institut Pasteur International Network (including at the Institut Pasteur in Paris). With its diverse expertise, including vaccine production at the Institut Pasteur in Dakar (a WHO pre-qualified yellow fever vaccine manufacturer), Tehran (childhood vaccines), and Tunis (rabies vaccine), the development of advanced vaccine candidates as described above and for whooping cough (Locht, Lille) and hand-foot-and-mouth disease (Huang, Shanghai), and surveillance, epidemiology, and academic research, the institutes of the Institut Pasteur International Network continue to drive innovation in vaccinology more than a century after Louis Pasteur's death.

WIDESPREAD EFFORTS TO COUNTER ANTIBIOTIC RESISTANCE

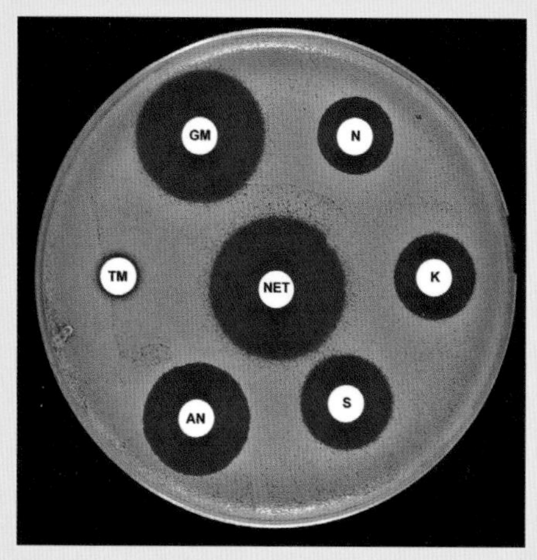

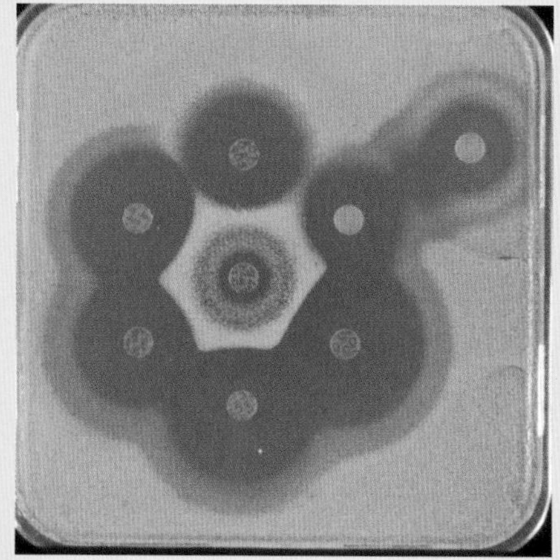

Antibiotics are highly effective drugs against bacteria and the diseases they cause, from mild conditions (ear infections, tonsillitis, etc.) to severe diseases such as meningitis and septicemia. Penicillin, discovered by Fleming in 1928, and the other families of antibiotics developed subsequently saved millions of lives. But as Fleming himself predicted, bacteria are capable of adapting and becoming resistant to these treatments, and today the efficacy of antibiotics is under threat. Some bacteria, known as multidrug-resistant bacteria, are now insensitive to several antibiotics. Antibiotic resistance is a problem in countries all over the world, but to varying degrees, often linked to the country's antibiotic consumption. Human medicine, veterinary medicine and environmental pollution by antibiotics are contributing to the emergence of a global phenomenon. In France, an "Antibiotic Plan" was launched in the early 2000s as part of efforts to preserve the efficacy of antibiotics. But despite some improvements, France remains one of the world's top consumers of antibiotics.

HOW DO BACTERIA BECOME RESISTANT?

Bacteria become resistant to antibiotics either because of a spontaneous mutation in their DNA or by acquiring resistance genes transferred from other bacteria. Taking prescribed antibiotics destroys all susceptible bacterial strains, effectively favoring non-susceptible strains and giving them free rein to proliferate. That is why prescribing antibiotics for no good reason merely selects the most resistant bacteria and promotes their circulation in populations.

In *Mycobacterium tuberculosis*, resistance is acquired by spontaneous chromosomal mutation, and then selection. Resistance to rifampicin occurs via a mutation in the DNA polymerase gene which enables the replication of genetic material. For other TB drugs such as isoniazid or ethionamide, the prodrug only becomes effective when activated by an enzyme in the bacillus. Resistance to these drugs involves a mutation that affects the gene coding for this enzyme.

Above left: Antibiogram of an aminoglycoside-resistant *Staphylococcus aureus*: the KmTm phenotype.

Above right: Culture of a strain of *Staphylococcus aureus*. Only the antibiotic in the central circle (erythromycin) is active; all the other antibiotics are inactive.

THE INSTITUT PASTEUR ON THE FRONT LINE OF THE FIGHT AGAINST ANTIBIOTIC RESISTANCE

The Institut Pasteur International Network, including the Institut Pasteur in Paris, is involved in all aspects of research into antibiotics and antibiotic resistance. Philippe Glaser, Head of the Ecology and Evolution of Antibiotic Resistance Unit at the Institut Pasteur, is coordinating the research program "Combating antibiotic resistance", involving 40 multidisciplinary teams from the Institut Pasteur International Network. The aim is to improve understanding of the molecular mechanisms underpinning resistance and to develop new anti-TB drugs and alternative therapeutic strategies.

The BIRDY project at the Institut Pasteur in Dakar in December 2014. The aim of this project was to study neonatal infections and antibiotic resistance in infants in low-income countries.

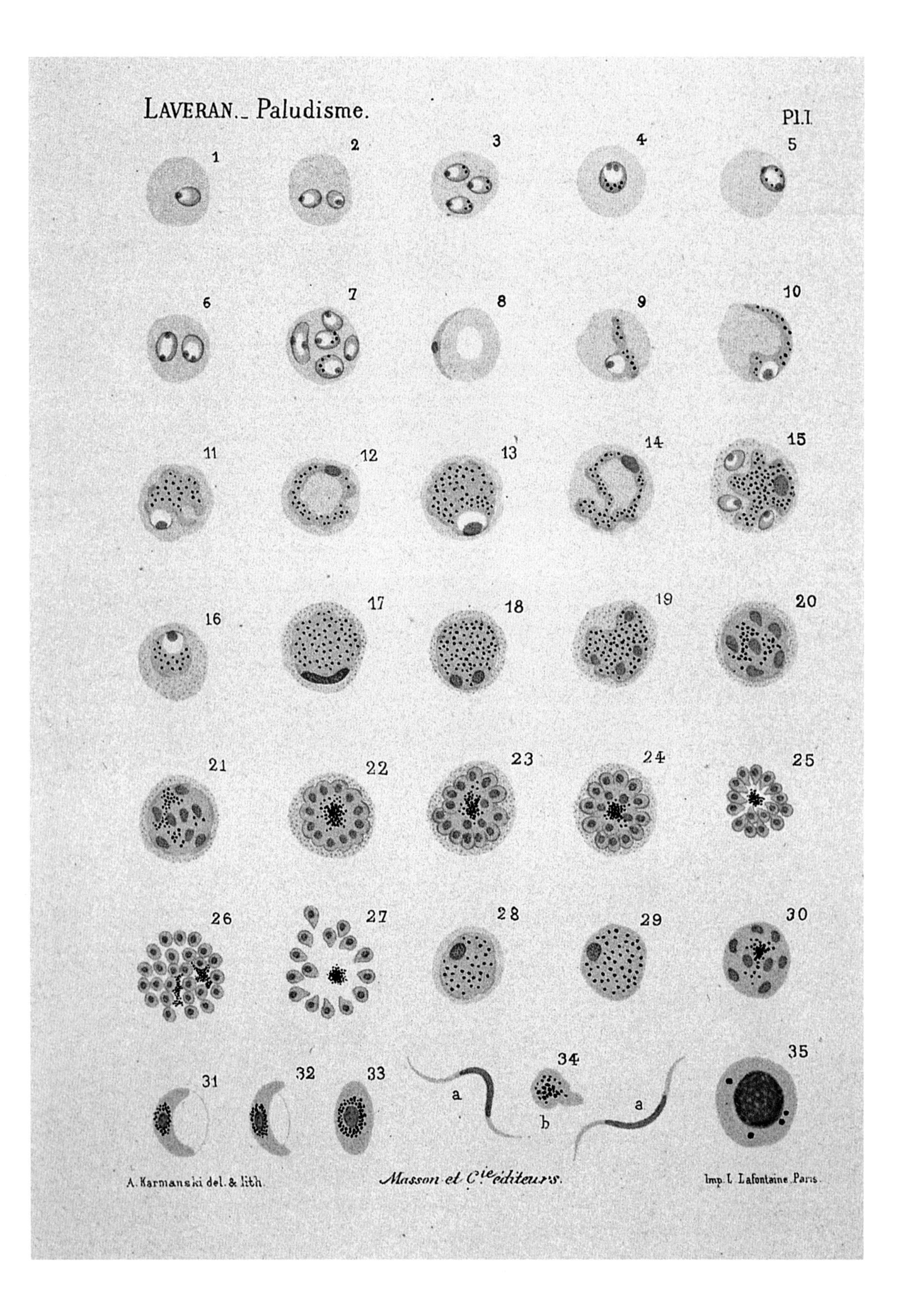
LAVERAN._ Paludisme.
Pl.I.
1
2
3
4
5
6
7
8
9
10
11
12
13
14
15
16
17
18
19
20
21
22
23
24
25
26
27
28
29
30
31
32
33
34
35
a
b
a
A. Karmanski del. & lith.
Masson et Cie éditeurs.
Imp. L. Lafontaine. Paris.

Malaria

and the Globalization of Vector-Borne Diseases

MALARIA
IN FIGURES

212 MILLION
cases each year

429,000
deaths, including 70% of children under 5

90%
of cases and 92% of deaths in Sub-Saharan Africa

Few diseases have had such a lasting impact on history and human life as malaria. This parasitic affliction, which occurs seasonally rather than in the form of outbreaks, takes its name from the Italian *mala'aria* meaning bad air. This term refers to the belief that stagnant marsh water can contaminate the air making it insalubrious. In ancient times, the parasite responsible for malaria (*Plasmodium*) was rife in the Mediterranean Basin. Hippocrates described fever spikes every third or fourth day (tertiary or quartan fever)—a rate that corresponds to the parasite life cycle in the human body. The fall of the Roman Empire and ensuing neglect of land around Rome—the *Campagna Romana*—exacerbated the malaria infestation and enabled *Plasmodium* to colonize Northern Europe before spreading to the New World with triangular trade.

Malaria almost certainly peaked in Europe between the 17th and 19th century, when Alphonse Laveran discovered the parasite (see page 64). From the 20th century onwards, the decline of the disease in Europe was linked to a combination of factors, including the draining of marshland, changes to farming methods and an increase in living standards. With a temperature rise of one or two degrees however, global warming could shatter the fragile *status quo* that we have enjoyed ever since.

Opposite: Research by Alphonse Laveran on the hematozoon (blood parasite) that causes malaria. Various stages as seen on fresh blood. Watercolor plate, 1880.

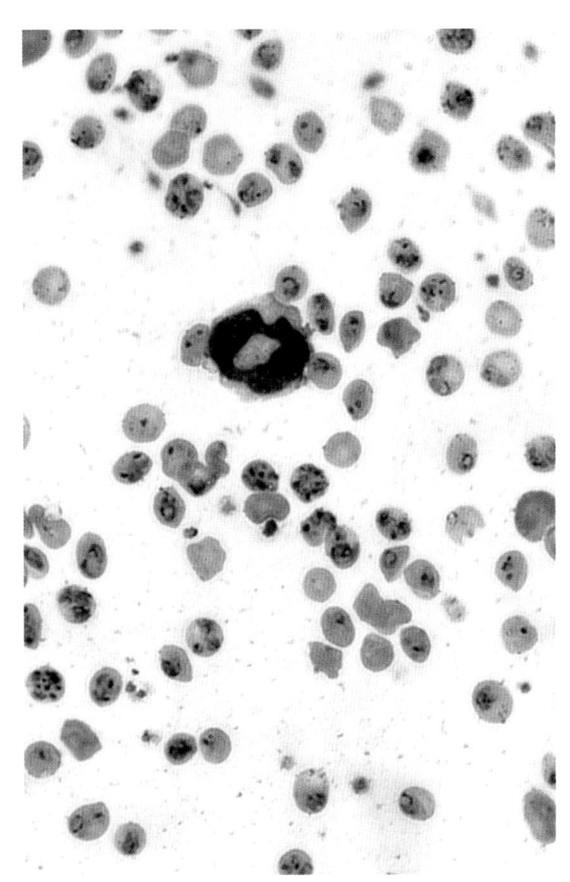

ALPHONSE LAVERAN, THE ARMY DOCTOR WHO DISCOVERED *PLASMODIUM*

A medical officer like his father, who ended his career as director of Val-de-Grâce military hospital, Alphonse Laveran (1845-1922) was appointed to Bône Hospital in Algeria in 1878, then to Biskra and Constantine. He was fascinated by malaria which became the focus of his research. In 1880, while examining blood from a patient under the microscope, he observed pigmented spherical bodies and other spindle-like elements which made a wave-like motion and were capable of displacing the neighboring red blood cells. Laveran was convinced that he had found the pathogenic agent responsible for malaria. And he had indeed isolated the *Plasmodium falciparum* protozoa, the cause of a scourge that has blighted humanity since the dawn of time. He sent a memo to the French Academy of Medicine and, four years later, published his *Traité des fièvres palustres* in which he suggested that the parasite was transmitted by mosquitoes. But there were reservations about this theory because scientists believed that malaria was caused by a bacterium and not this strange coil-like organism. Laveran was appointed Professor of Military Hygiene at Val-de-Grâce in 1884 and was rewarded for his tenacity with the Nobel Prize in Physiology or Medicine in 1907. He had studied at the Institut Pasteur in 1889 and joined the institute eight years later. Following his Nobel Prize, he donated 100,000 francs to convert and equip the tropical disease laboratory. He then carried out research into parasitic diseases in these modernized facilities until he died in 1922.

Top left: *In vitro* culture of mouse red blood cells infected with *Plasmodium chabaudi*, the malaria agent in rodents, transmitted by *Anopheles stephensi*. *P. chabaudi* is used as a model for human malaria. In the center, a white blood cell.

Top right: Alphonse Laveran (1845-1922).

A heavy burden on the African continent

In 2015, the World Health Organization (WHO) estimated that half of the world's population was at risk of contracting malaria. That year, there were 212 million sufferers, 429,000 of whom died. The infection is particularly severe in children under five, pregnant women, immunodeficient patients and, to a lesser extent, in people who have never been exposed to it and are likely to be, for example non-immune migrants, homeless populations and travelers.

Although South-East Asia, Latin America and the Middle East have not been spared, the African continent is the most affected—90% of cases and 92% of deaths occur in Sub-Saharan Africa. Recent data however provides a glimmer of hope. Better access to preventive measures, diagnosis and curative treatment over the last few years is starting to pay off. According to WHO, the mortality rate fell by 29% between 2010 and 2015, though the goal to reduce the incidence of malaria by 40% by 2020 seems unattainable. And vigilance is still necessary as there is growing resistance to antimalarial drugs, and outbreaks have occurred outside of traditional endemic areas, particularly in South Africa, Costa Rica, Venezuela and Malaysia.

In France, the number of imported cases was estimated at 4,735 in 2016. There were two suspected indigenous cases where mosquitoes were thought to have infected the patients on French soil. Generally, these observations are reported at airports where mosquitoes can arrive with freight or luggage.

A potentially fatal disease

Malaria is caused by a parasite of the *Plasmodium* genus. Symptoms include the sudden onset of a high fever, 8 to 30 days after the infectious bite. Patients may also suffer chills, headaches, weakness, muscle pain, digestive problems (vomiting, diarrhea) and coughing. Malarial attacks, which consist of cold, hot and sweating stages, are followed by periods of respite. This frequency corresponds to the *Plasmodium* life cycle, as the attacks coincide with parasite multiplication and red blood cell rupture.

Parasites, mosquitoes and humans—a triangular relationship

Malaria is a vector-borne disease, meaning that the pathogenic agent infects humans (or animals) via living organisms that host it temporarily, most often blood-sucking insects. *Plasmodium* is spread by female *Anopheles*, a mosquito genus with 400 recognized species, thirty of which are likely to transmit the parasite. These mosquitoes need blood to reproduce and mature their eggs, which are then laid in water and hatch into larvae before becoming adults. Females are infected by biting individuals with malaria and human-to-human transmission occurs via mosquitoes. Direct person-to-person transmission can happen when the parasite passes from mother to child by crossing the placental barrier, or following a blood transfusion with infected blood.

Gametocytes: cells that are widespread in the blood of individuals with malaria and the precursors of malaria parasite gametes (reproductive cells).

Plasmodium has a complex life cycle. The parasites multiply by simple cell division (asexual stage) in the liver and red blood cells of the human host. Some of them then undergo metamorphosis and male and female sexual parasites are formed. When a female *Anopheles* mosquito bites an infected individual, it sucks up gametocytes which reproduce in its stomach and then reach its salivary glands. The mosquito can then bite and infect a human and the *Plasmodium* life cycle continues.

FIVE *PLASMODIUM* SPECIES INFECTING HUMANS

There are five parasite species that infect humans. *Plasmodium falciparum* is responsible for the most common and severe forms of malaria. It is the dominant species in Africa and can also be found in the tropical regions of Latin America and Asia. If not treated within 24 hours, the disease it causes can be fatal, particularly due to the severe anemia caused by a drop in hemoglobin levels. The parasited red blood cells can also adhere to each other and to the wall of small blood vessels carrying blood to the brain, causing poor-prognosis cerebral malaria. The less harmful *P. vivax* coexists in most regions where *P. falciparum* is prevalent and predominates in South America. The two other, rarer species can result in late relapses—*P. ovale*, present in West Africa, can cause attacks four to five years after the first symptoms, and *P. malariae*, which is very unevenly distributed throughout the world, can flare up as late as 20 years after an initial episode. Human cases of infection by *P. knowlesi*, which usually infects monkeys, are also regularly reported.

Opposite top: Immunofluorescence image revealing the Pf155/RESA protein in red blood cells infected by *Plasmodium falciparum*.

Opposite bottom: An *Anopheles darlingi* larva bred at the Institut Pasteur Vectopole in French Guiana. *Anopheles darlingi* is the main malaria vector in French Guiana.

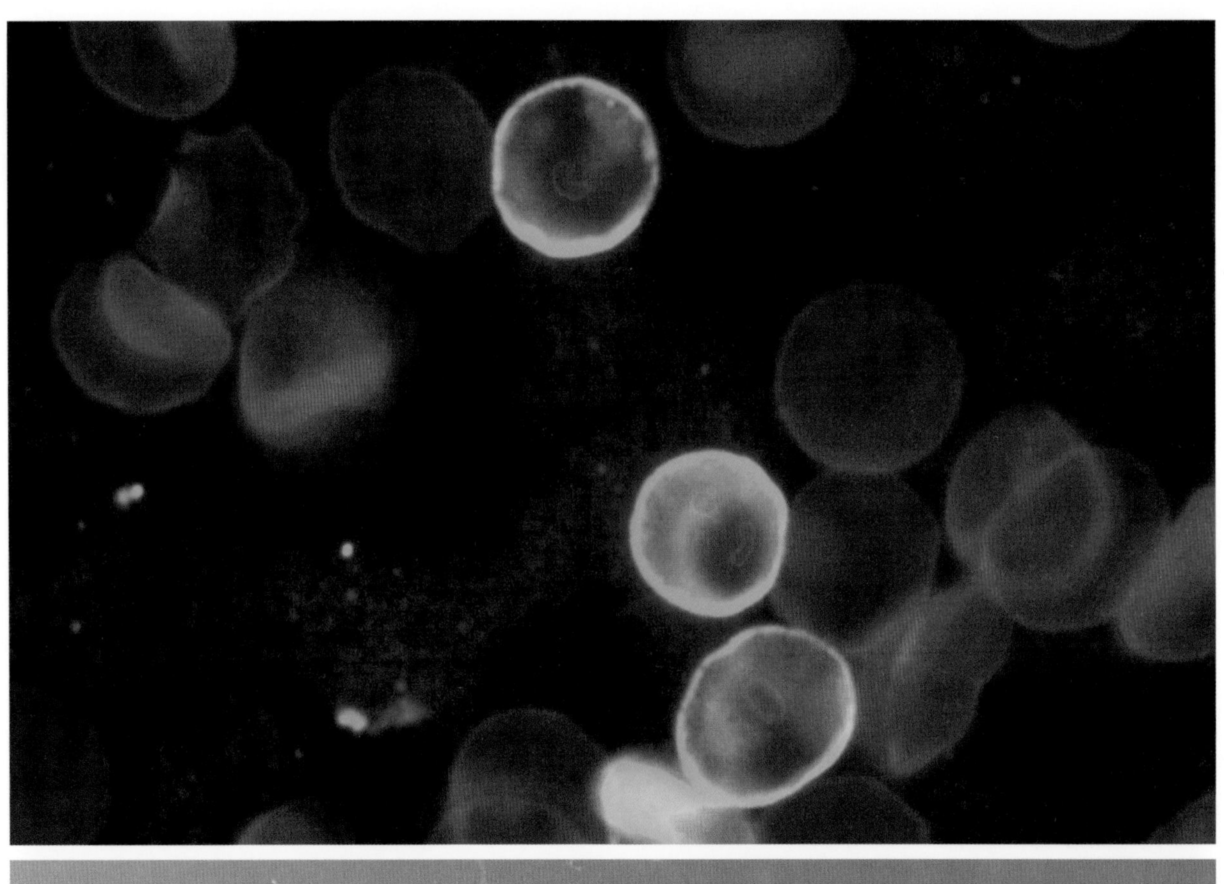

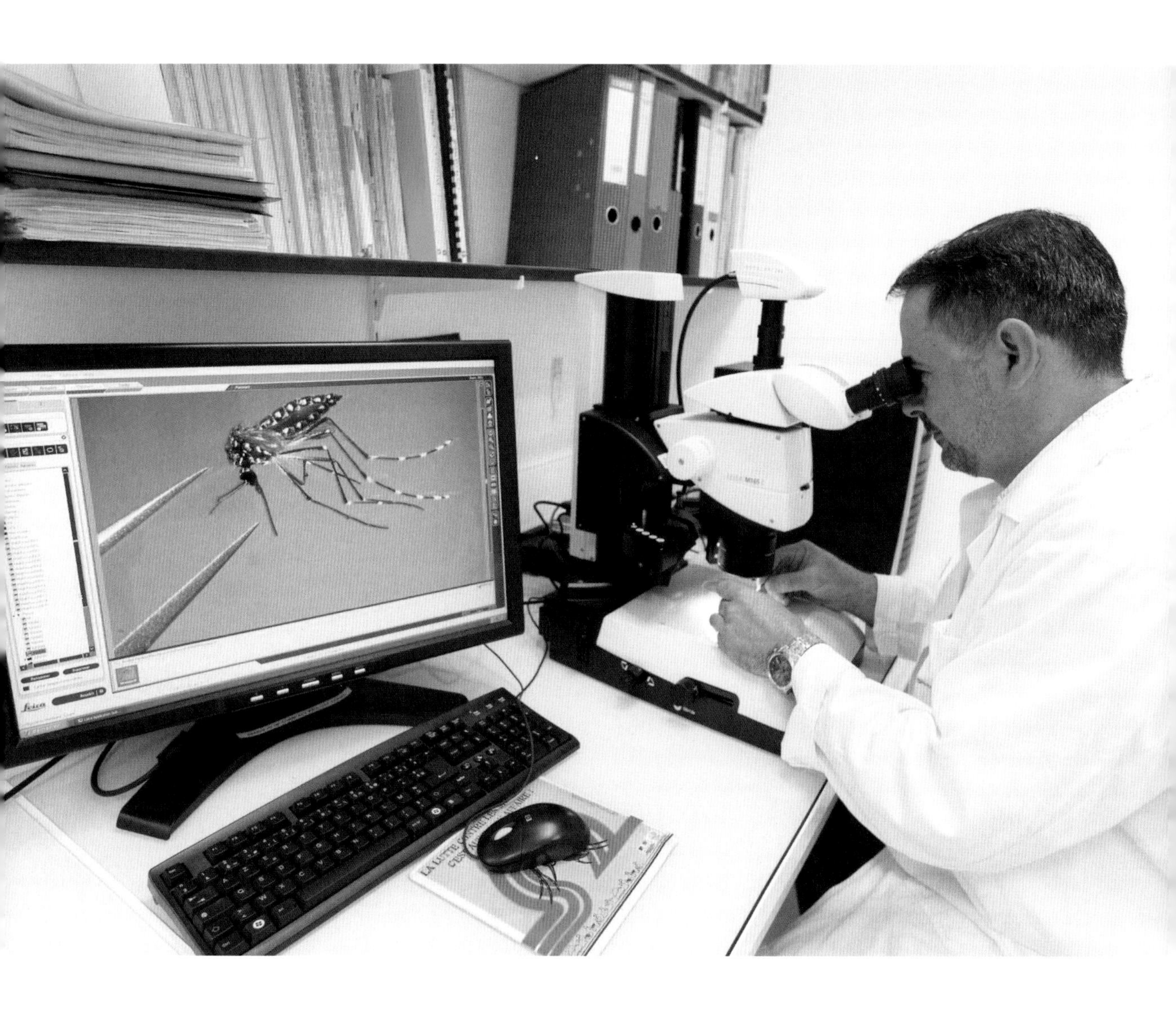

Prevention and treatment: escalation of resistance

There are two types of treatment—preventive treatment to avoid contamination, and curative treatment to manage malarial attacks. Although doctors can prescribe several drugs in each case, these treatments are facing growing parasite resistance.

A technician observing a mosquito under a binocular microscope in the Medical Entomology Unit at the Institut Pasteur in French Guiana. Taxonomy and systematics are key research areas in the Medical Entomology Unit.

"The Institut Pasteur had the foresight to conduct clinical, epidemiological and genomic studies to implement a strategy leading to the discovery of the K13 gene's involvement in artemisinin resistance."

Didier Ménard, Head of the Malaria Molecular Epidemiology Unit at the Institut Pasteur in Cambodia

Preventive treatment

First-line preventive treatment does not however involve drugs but measures likely to stop mosquito bites. Large-scale use of these vector control strategies has reduced the incidence of malaria over the last few years in highly endemic areas. The WHO and local health authorities recommend two particularly effective courses of action—using insecticide-impregnated mosquito nets and spraying residual insecticides indoors. Travelers can reduce the risk of getting bitten by using mosquito nets, wearing long, insecticide-impregnated clothing, and applying insect repellents to exposed skin.

Preventive treatment is necessary when traveling to malaria-endemic areas, particularly for pregnant women and children, who have an increased risk of severe malarial attacks. This chemoprophylaxis is only available on medical prescription, as the type and length of treatment depend on several factors—the destination, individual situation (age, medical history, drug intolerance, etc.) and length of the trip. As well as the general advice available online, experts from the Institut Pasteur Medical Center offer specialist consultations to recommend steps for each and every situation. Four main drugs are used. The choice is determined by whether resistance exists in the region

Anopheles stephensi, mosquitoes produced in the insectarium at the CEPIA (Center for the Production and Infection of Anopheles) for various research programs and teams. *Anopheles stephensi* is one of the *Anopheles* species responsible for the spread of malaria in Asia.

visited: chloroquine (in increasingly rare areas that are still sensitive), atovaquone-proguanil, mefloquine (which has tolerance issues) and doxycycline.

Chemoprophylaxis is not reserved exclusively for travelers, as since 2012, WHO recommends seasonal preventive medication in Sub-Saharan Africa for pregnant women and children under five.

Curative treatment

Treatment begins following confirmation of the diagnosis, which usually involves examining a blood sample for *Plasmodium* genus parasites under the microscope. Rapid diagnostic tests have been widely available in recent years and can be used to obtain a result in under 30 minutes by collecting a single drop of blood from a finger tip. In July 2016, the Institut Pasteur in Bangui launched the PaluFlag study with the company Horiba to develop more straightforward, cheaper tests for diagnosing malaria as quickly as possible. It also involves assessing the proportion and type of non-malaria fevers in the Central African Republic (CAR).

Plasmodium resistance to conventional treatments, such as chloroquine and the sulfadoxine/pyrimethamine combination, became widespread in the 1950s and 1960s. The emergence of resistance led to the WHO recommending artemisinin-based combination therapy, or ACT, in Southern Asia then Africa as early as 2001. Artemisinin is derived from a plant that has been known for its medicinal properties in China for over 2,000 years. This recommendation is still in place as ACT has proved effective despite resistance to artemisinin.

KARMA: GLOBAL MAPPING OF ARTEMISININ RESISTANCE

With the help of 41 partners, including 13 members of the Institut Pasteur International Network, the consortium has mapped global artemisinin resistance, linked to Kelch gene mutations on chromosome 13 of *Plasmodium falciparum* (K13). This molecular marker, discovered in 2014 by Didier Ménard's Malaria Molecular Epidemiology Unit (Institut Pasteur in Cambodia), can be used to detect artemisinin resistance using a quick and simple test. KARMA has demonstrated that parasite resistance to artemisinin derivatives, which first emerged in Cambodia, remains confined to South-East Asia. These findings need to be put into perspective given the frequent use of ACT in this region and the movement of populations (particularly rubber plantation workers). For the time being, the African continent seems to have been spared. To counter this resistance and its geographical spread, WHO is investigating the option of using lengthier treatments, triple combination therapy and several different drugs at the same time in the same country.

Resistance epidemiology

Growing parasite resistance to available treatments poses a major public health problem. Firstly, it requires regular assessment of *Plasmodium* sensitivity and evolution in the different regions of the world. The WHO is involved in this task alongside the Institut Pasteur International Network (IPIN). Institutes in malaria-endemic countries in particular, like Cambodia, Madagascar, Senegal, Côte d'Ivoire, Niger and the Central African Republic, are especially active. As for the Institut Pasteur in French Guiana, it is a National Reference Center for Malaria Chemoresistance in the Antilles-French Guiana region. The team led by Lise Musset is studying parasites and their resistance by testing the sensitivity of collected samples to eight molecules used in therapies, and by looking for resistance genes. It has shown that a mutation of the pvmdr1 gene confers chloroquine resistance in *Plasmodium vivax* strains circulating in French Guiana.

The situation is, however, less alarming than in Cambodia where Institut Pasteur researchers are coordinating the K13 Artemisinin Resistance Multicenter Assessment Consortium (KARMA).

A mosquito trap is set up in the town of Matoury as part of research on malaria transmission during an entomology field trip to Maripasoula, French Guiana, in March 2016.

As well as this monitoring activity, the development of new treatments may help to overcome, albeit temporarily, the resistance issue. This is the goal of the MaPI (Malaria Protease Inhibition) project, coordinated by Jean-Christophe Barale and his Biology of Malaria Targets and Antimalarials team at the Institut Pasteur in Paris. With the backing of the French National Research Agency and in partnership with Sanofi, this program focuses on two new therapeutic targets—the SUB1 and SUB2 proteins, identified in 2012 and used by *Plasmodium* to enter and exit liver and red blood cells.

Barriers to vaccination

An initial malaria vaccine, RTS,S/ASO1, received a favorable opinion from the European Medicines Agency and has been available since 2015. Its efficacy is, however, limited and it is reserved for children in their early years. In addition, the four injections required do not confer lasting immunity. The WHO launched a pilot study in three Sub-Saharan African countries to test its ability to reduce child mortality linked to malaria.

The main challenge facing the *Plasmodium* vaccine is that the parasite evades the immune system. An initial attack does not lead to antibodies being produced or memory of infection. Only individuals with chronic malaria have some degree of protection. To develop this vaccine, researchers fused a parasite antigen to the hepatitis B antigen to trigger the immune response.

Salaheddine Mécheri's Biology of Host-Parasite Interactions Unit at the Institut Pasteur has developed another vaccine strategy with Robert Ménard from the Malaria Infection & Immunity Unit. The scientists isolated a parasite gene that codes for the HRF (Histamine Releasing Factor) protein, which appears to play a key role in the early stages of infection. When injected into mice, a *Plasmodium* parasite lacking this gene showed delayed development in the liver cells. Significant production of cytokine IL-6, known to stimulate the immune response against intruders, has also been observed. So, the organism defends itself against the parasite and retains lasting immune memory of the attack. This opens up possibilities for developing a vaccine using genetically modified live parasites.

Opposite: Mosquito larvae are collected in Cacao, French Guiana, by an Institut Pasteur scientist during the Zika virus outbreak in March 2016.

Entomology to fight vector-borne diseases

Over 3 million insect species have been identified to date, which highlights the immensity of the task facing entomologists, the scientists that study these living beings. A certain number of species are likely to transmit infectious agents to humans and are disease vectors. These are biting insects, which are often partial to blood meals. This worrying bestiary features a whole host of insects—ticks (Lyme disease), flies (sleeping sickness or trypanosomiasis for the tsetse fly), blackflies (onchocerciasis), fleas (plague), bugs (Chagas disease) and, of course, mosquitoes that transmit infectious and parasitic diseases, including malaria, chikungunya, dengue, Zika, and also Rift Valley fever, yellow fever, West Nile fever, Japanese encephalitis or lymphatic filariasis. As they are capable of spreading over one hundred diseases, mosquitoes are public enemy number one.

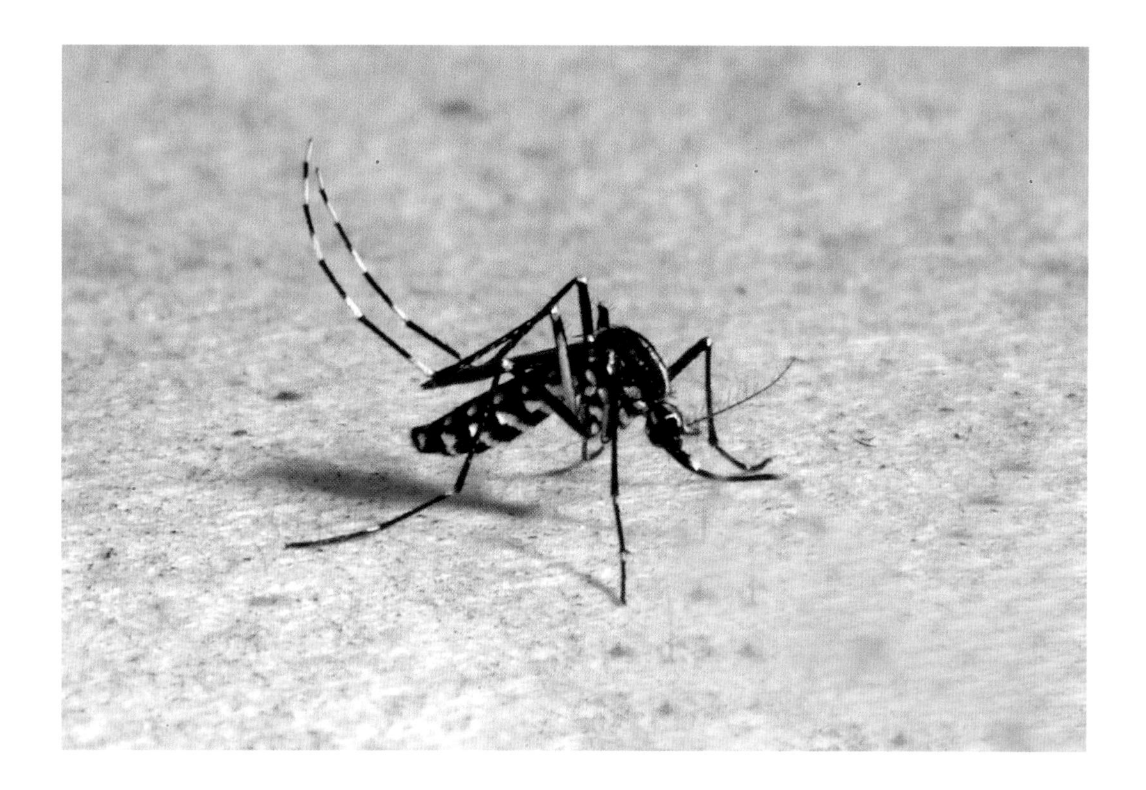

A female *Anopheles darlingi* mosquito bred at the Institut Pasteur Vectopole in French Guiana.

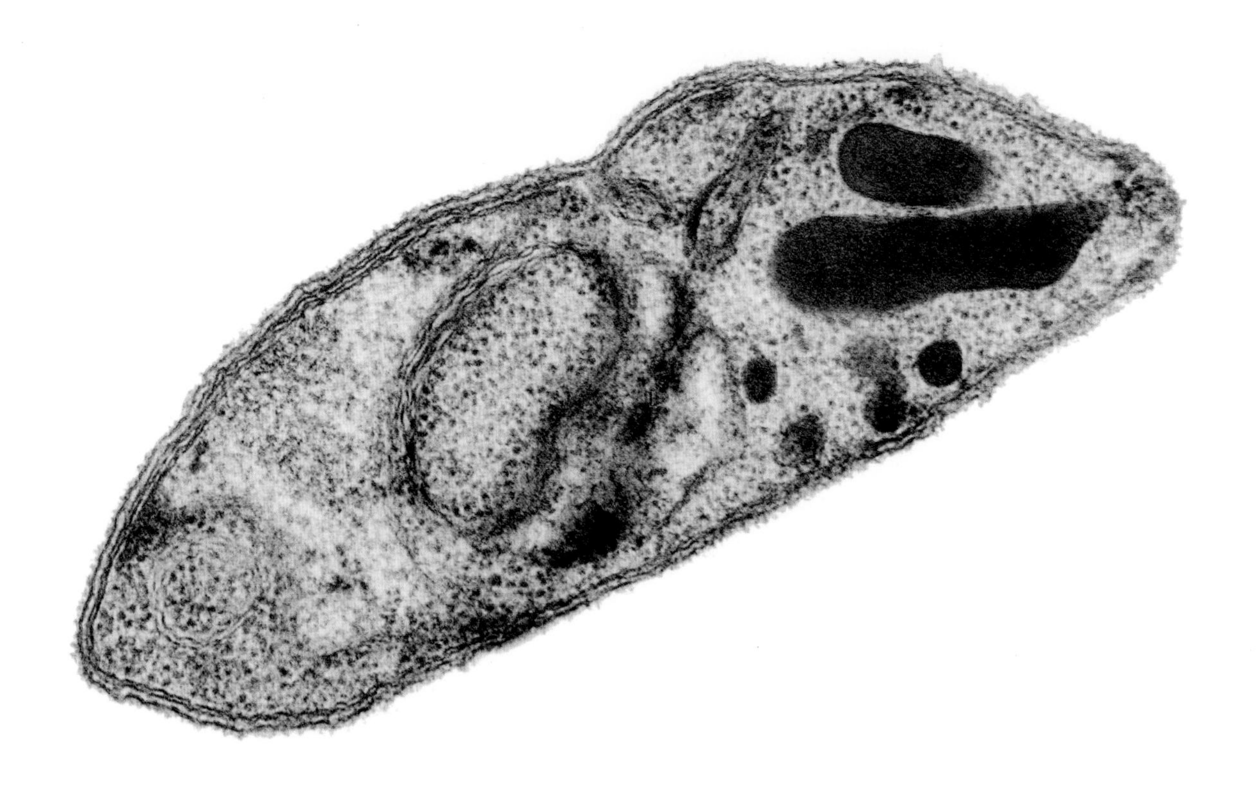

Understanding their physiology and ability to adapt is key in the fight against vector-borne diseases, especially for mosquitoes that spread the dengue, Zika or chikungunya arboviruses. At the Institut Pasteur in Paris, the Arboviruses and Insect Vectors Unit directed by Anna-Bella Failloux is studying interactions between mosquitoes, viruses and the environment. It seeks to identify the factors underlying the emergence of pathogens responsible for human outbreaks—factors linked to the interaction between arboviruses and their vectors, and also to environmental and/or climate changes that affect the organization of mosquito populations. The unit has mosquito breeding facilities, which house the most harmful species for humankind, and a BSL-3 laboratory to carry out experimental infections in animal models. Anna-Bella Failloux and her team work closely with IPIN member institutes.

The merozoite stage of *Plasmodium falciparum*, the infectious form that invades red blood cells.

The Institut Pasteur in French Guiana has launched an extensive epidemiological survey (EPI-ARBO) among the population to accurately assess the impact of arboviruses in the country. Since 2014, it has been home to an insect vector research facility—the Vectopole Amazonien Émile-Abonnenc, which is named after the first entomologist of the Guianese institute. In this insectarium, entomologists classify the mosquitoes which, with a cycle of approximately 700 generations per year, continually evolve producing new species. For the moment, no fewer than 250 of them, out of the 3,500 species known worldwide, have been identified in this region of South America. A laboratory is monitoring the development of arthropod resistance to insecticides. Also, like in Paris, a BSL-3 laboratory is carrying out experimental contaminations. In Cambodia, a medical entomology facility headed by Sébastien Boyer and supervised by Didier Fontenille, director of the local Institut Pasteur, has been in place since 2015. As well as malaria, dengue, Zika and chikungunya, Japanese encephalitis is also the focus of in-depth research. Other institutes, like those in Bangui or Dakar, are actively involved in monitoring mosquito vectors of arboviruses and have formed a unique international network.

Mosquitoes being sorted in the Medical Entomology Unit at the Institut Pasteur in French Guiana.

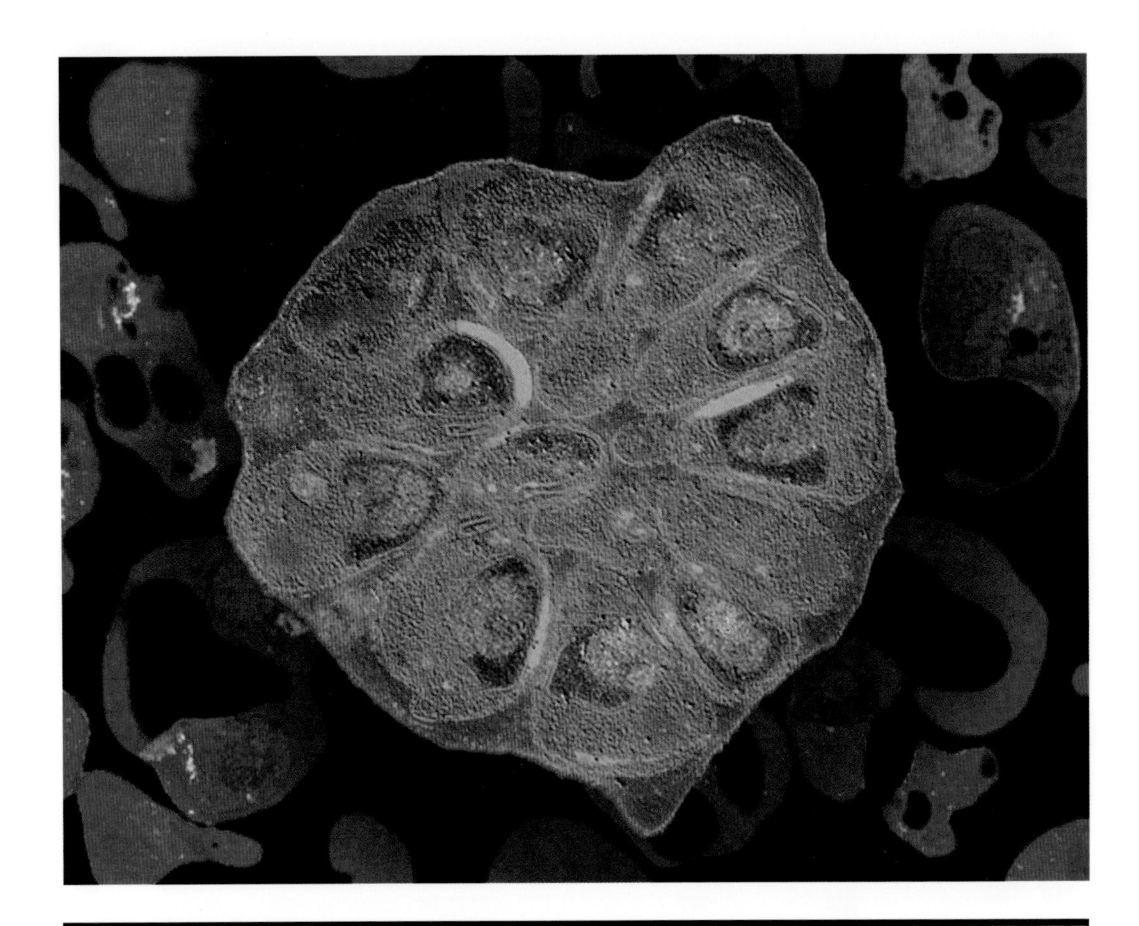

WHEN GENES AND PARASITES WORK TOGETHER

Hereditary anemia—mainly sickle-cell and thalassemia—includes hemoglobin disorders of genetic origin. As early as the 1920s, doctors wondered whether there was a link between these diseases and malaria. Firstly, they had noticed that the geographical distribution of the diseases overlapped, both in Sub-Saharan Africa and the Mediterranean Basin. Secondly, an explanation was needed as to why the genes of these harmful diseases were so widespread. Finally, Italian specialists had noted than in Sardinia, which at the time was a high malaria-endemic region, thalassemia was common in coastal areas but practically non-existent in the mountains. A population geneticist, John B. S. Haldane, put forward a solution to this riddle—those who carry one abnormal copy of the gene (heterozygotes) are more resistant to malaria than those with normal hemoglobin. In heterozygotes, thalassemia provides a selective advantage in malaria-endemic areas as defective red blood cells hinder the parasite life cycle. Hereditary anemia is, however, rare in malaria-free areas, for instance in the mountains where mosquitoes disappear above a certain altitude. This theory was confirmed by Anthony C. Allison in 1954. He demonstrated that plasmodium levels in the blood were lower in people living on the banks of Lake Victoria as they were heterozygotes for sickle-cell anemia. So, an inherited genetic disorder can interfere with a vector-borne parasitic disease, which itself selects human populations.

Artificially colored transmission electron micrograph showing a red blood cell infected with *Plasmodium falciparum*. The infected erythrocyte is shown in red, and the parasite is shown in purple with nuclei in blue.

AIDS

Research Uniting Disciplines

AIDS ACROSS
THE GLOBE IN 2017

37 MILLION
people living with HIV

17.5 MILLION
people on triple therapy (84% more than in 2010)

1.5 MILLION
deaths due to AIDS (42% fewer than in 2004)

"The greatest health catastrophe in human history". These were the words used by WHO to describe the AIDS pandemic, based on the statistics available since the 1980s. Nearly 35 years on from the discovery of HIV-1 (human immunodeficiency virus-1) by Institut Pasteur scientists, 37 million people worldwide live with the virus. Fewer than half of them (17.5 million) have access to triple antiretroviral therapy, and approximately 1.5 million people die of AIDS each year. In France figures have remained stable since 2011, with almost 6,000 new cases in 2015.

Despite appearances, this cloud has a silver lining. For several years the trend has been reversing: annual deaths have fallen by 42% since 2004 and new HIV infections by 35% since 2000; access to treatment has increased by 84% since 2010. The number of HIV cases has never been higher, but this is because more people are seeking diagnosis and being treated. The rate of spread has declined following coordinated international response. With realistic ambitions and clear objectives, UNAIDS (the Joint United Nations Program on HIV/AIDS) is now able to forecast an end to the epidemic by 2030.

Opposite: Viral particles of HIV-1, the causative agent of AIDS, in 1986. Virus isolated in 1983 by Luc Montagnier, Jean-Claude Chermann and Françoise Barré-Sinoussi.

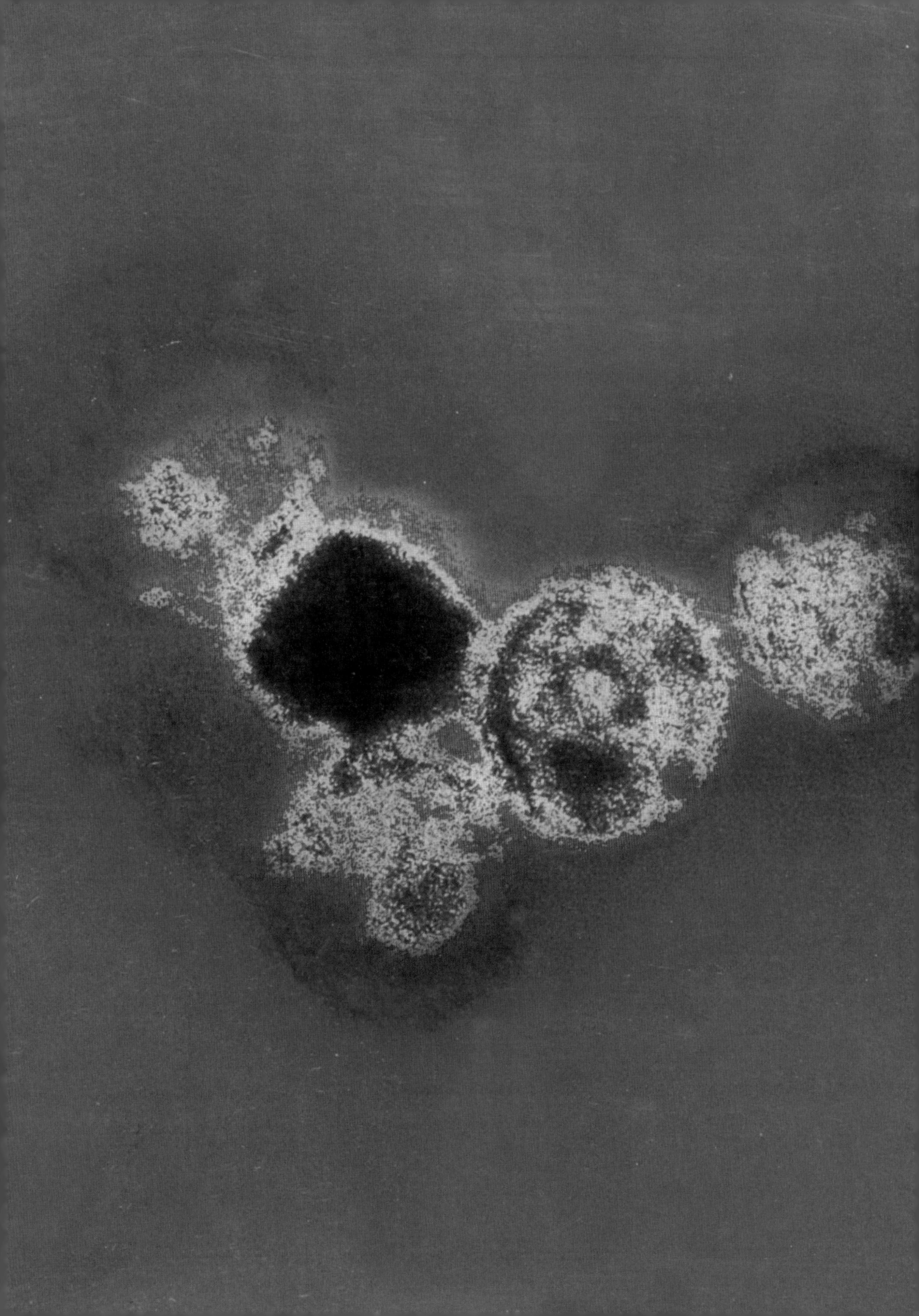

VIRUSES AND LIVING ORGANISMS

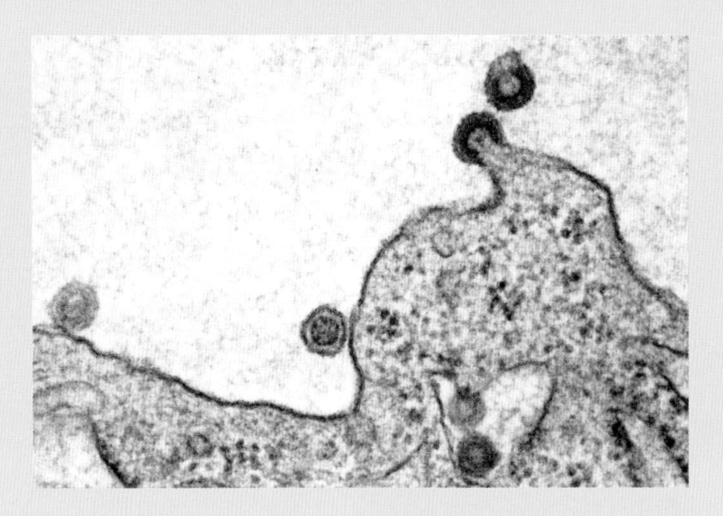

The term "virus" is Latin for "poison", and was used indiscriminately by Louis Pasteur to name various pathogens. He applied the term "virus" to anthrax, a bacillus with which he was very familiar, and spoke in equal measure of the rabies virus, which evaded his microscope. He knew that the rabies virus was present in the brain and saliva of patients, and he could inoculate animals with it, but he was unsuccessful in observing or culturing it. At the end of the 19th century, microbiologists had understood that certain diseases are caused by agents far smaller than bacteria. These agents are known as "ultra-filterable", since they pass through even a porcelain filter—the famous Chamberland filter.
The size of these microorganisms—ranging from 10 to 400 nanometers—means that viruses are, on average, 1,000 times smaller than bacteria. With ultracentrifugation, electron microscopy and X-ray diffraction techniques, knowledge of their structure vastly improved in the mid-20th century. It became progressively clear that a single, isolated virus is not capable of any of the activities that characterize bacteria and, more generally, life: it neither metabolizes nor grows, nor does it move, reproduce or multiply. A virus is an absolute parasite, composed of genetic material and structural proteins, but can only replicate by taking over a living cell and subverting its cellular machinery for its own purposes. This gives rise to the question that has long divided biologists: are viruses living beings?
Virology was established as a science in the 1950s, and in 1957 the Institut Pasteur's André Lwoff gave his own definition of a virus:
a virus possesses only one type of nucleic acid (RNA or DNA);
- its nucleic acid is its only means of replication;
- it does not grow or divide, and multiplies in the cell via synthesis and assembly;
- it contains no information on intermediary metabolism enzymes;
- it multiplies only inside cells.

As this definition suggests, we can make a distinction between DNA viruses, such as chicken pox or papillomavirus—which is implicated in cervical cancer—, and RNA viruses, such as influenza or HIV. Their genome may be single stranded or double stranded. It is surrounded by a protein structure called a capsid. Some viruses, known as "enveloped viruses", have an additional envelope, which is the outermost structure.
Humans are not the only living beings to suffer the ill effects of viral invasion. In 1917, the Institut Pasteur scientist Félix d'Hérelle described ultraviruses—another name for ultra-filterable viruses—, which attacked bacteria: these are known as bacteriophages. Although his own attempts failed, the use of bacteriophages in the treatment of bacterial diseases—or phage therapy—now shows promise as an alternative to antibiotics at a time when antibiotic resistance is on the rise. The team led by Laurent Debarbieux, within the Bacteriophage-Bacteria Interactions group at the Institut Pasteur, is conducting research in this field.

HIV budding at the surface of a thymocyte.

A changing virus

1

RNA: in the cell nucleus, production of a protein from a gene (DNA fragment) involves an intermediate phase: synthesis of RNA, a molecule with a structure very similar to DNA.

DNA: an essential component of chromosomes that carries genetic information. DNA is present in all human cells except red blood cells.

AIDS is caused by human immunodeficiency virus-1, which is contracted through sexual contact, contaminated blood and mother-to-child transmission. It belongs to the retrovirus family that contains RNA as its genetic material. The particular feature of retroviruses is that they have an enzyme called reverse transcriptase, which transcribes viral RNA into complementary DNA.

During transcription, a number of errors are produced and since HIV has no proofreading mechanism for identifying or correcting them, the virus is prone to mutations. However, what at first glance appears to be a defect is actually an advantage for the virus, since variant viruses are produced in a constant stream, and any new defense strategy used by the body is likely to come up against a resistant variant, which will then be selected. The virus' weakness is therefore its strength.

AIDS is an immune system disease. HIV attacks CD4+ T lymphocytes—immune system cells with a CD4 surface marker. Like all viruses, it subverts cellular processes to replicate and produce new viral particles. It can also remain dormant in infected cells and integrate its genetic material into their genome. The destruction of CD4+ T lymphocytes and defects in immune cell activation hinder the host's immune response, increasing susceptibility to opportunistic infections and certain cancers.

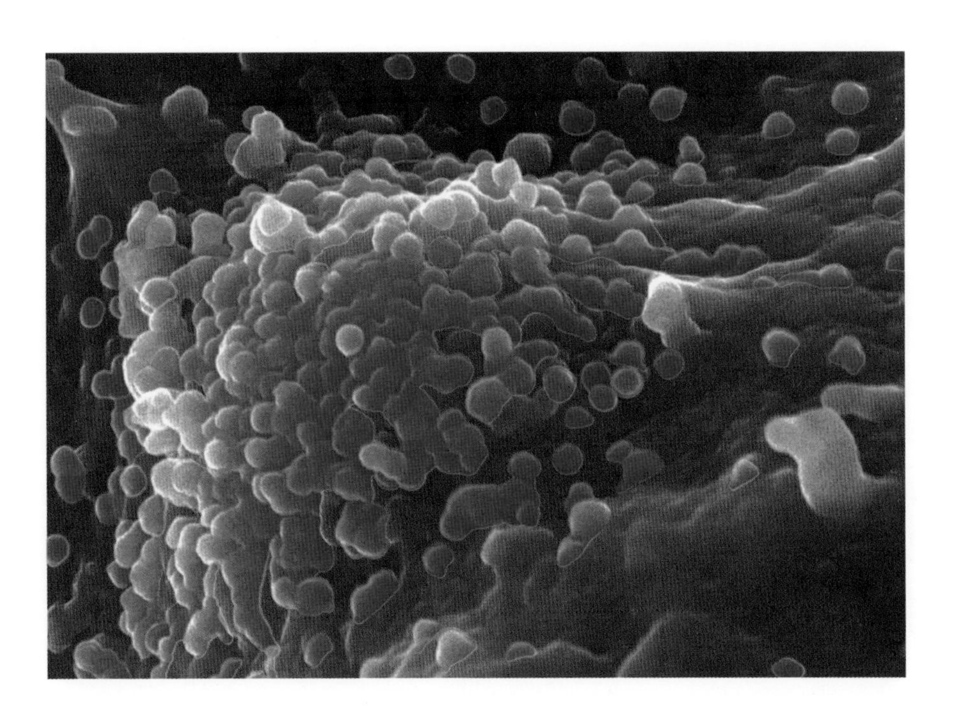

HIV particles budding at the surface of a lymphocyte.

HIV DISCOVERY

IN KEEPING WITH PASTEURIAN TRADITION

In 1981, when a "new disease" was reported in the USA, Robert Gallo (in Bethesda) had recently isolated the first human retrovirus responsible for a specific form of leukemia—HTLV-1 (human T-leukemia virus). At the Institut Pasteur, the Department of Virology had been focusing on oncogenic viruses under the directorship of Luc Montagnier since 1972. He was later joined by Jean-Claude Chermann and Françoise Barré-Sinoussi, who benefited from a legacy of outstanding virology research and the Institut Pasteur's contribution to the molecular biology revolution. The team had access to the best tools for detecting and assessing the activity of reverse transcriptase—the enzyme that converts HIV NRA into DNA and identifies its presence.

From 1982 onwards, the Institut Pasteur scientists formed a multidisciplinary group with virologists, immunologists and hospital clinicians, including Willy Rozenbaum (from Pitié-Salpêtrière Hospital). In 1983, Rozenbaum provided a lymph node biopsy from a 33 year-old patient suffering from generalized lymphadenopathy associated with immunodeficiency. Françoise Barré-Sinoussi detected minor reverse transcriptase activity in the cultured sample. In the days that followed, the signal grew stronger and clearer, and then began to decrease as the CD4 T lymphocytes disappeared. As an urgent measure and to avoid losing the virus, white blood cells from a healthy donor, provided by the Institut Pasteur Hospital blood service, were added to the culture medium. Reverse transcriptase production immediately resumed.
From that moment on, the team began to suspect that the causative agent of AIDS—which destroys CD4 lymphocytes—was not HTLV-1, which tends to render cells "immortal" as seen in leukemia. The scientists established a link with a feline retrovirus responsible for leukemia in cats, causing immunodeficiency that mostly proves fatal. Two further developments strengthened their conviction: electron microscopy images produced at the Institut Pasteur on February 4, 1983 by Jean-Claude Dauguet showed morphological differences with HTLV-1, and the specific HTLV-1 antibodies sent by Robert Gallo did not react with the virus isolated in Paris.
The first description of the virus—which the team called LAV (lymphadenopathy-associated virus)—was published on May 20, 1983 in the journal *Science*. It was only much later—in May 1986—that it was renamed HIV (human immunodeficiency virus). In 2008, the Nobel Prize in Physiology or Medicine was awarded to Luc Montagnier

> "We discovered HIV after our hospital colleagues approached us with questions about the causative agent of AIDS. Nothing would have been possible without multidisciplinary efforts and strong interaction with clinicians and patient representatives."

Françoise Barré-Sinoussi, laureate of Nobel Prize in Physiology or Medicine 2008

Opposite top: HIV researchers at the Institut Pasteur in 1985, including Prof. Luc Montagnier, Françoise Barré-Sinoussi (both 2008 laureates of the Nobel Prize in Physiology or Medicine) and Jean-Claude Chermann; Marc Alizon, Simon Wain-Hobson, Pierre Sonigo and Olivier Danos, who sequenced the HIV-1 virus in 1985 (with Stewart T. Cole, missing in the photo); Charles Dauguet, head of electron microscopy; Ara Hovanessian; and François Clavel, co-author of the article on HIV-2 sequencing in 1987 with Mireille Guyader, Mickael Emerman, Pierre Sonigo and Marc Alizon.

and Françoise Barré-Sinoussi for their discovery, jointly with the German virologist Harald zur Hausen for his identification of papillomavirus, which has been implicated in cervical cancer.

Over the next few years, the Institut Pasteur orchestrated or participated in a number of advances in the field of AIDS: it demonstrated transmission from mother to child, and from donors to recipients of blood or blood products; it also identified the CD4 receptor and HIV-2, and developed specific tests from 1985 onwards via a partnership with Sanofi Diagnostics Pasteur, to name a few. The discovery of LAV marked the start of a four-team race to sequence its genome. The Institut Pasteur group, coordinated by Simon Wain-Hobson, succeeded in providing the first nucleotide sequencing in 1985. This represented a crucial step in the development of a therapeutic strategy, for two reasons. Firstly, knowledge of viral genes shed light on the role of the ten viral proteins and replication mechanisms. Using these data, therapeutic targets were identified—representing an initial step towards the development of new drugs. Secondly, genome analysis rapidly showed the enormous variability of the virus and its propensity to produce therapy-resistant variants. This period also highlighted the need for several active substances effective on different viral proteins, and scientific concepts and principles for effective HIV treatment were in place before the end of the 1980s. However, it was not until 1996 that patients received triple therapy for the first time.

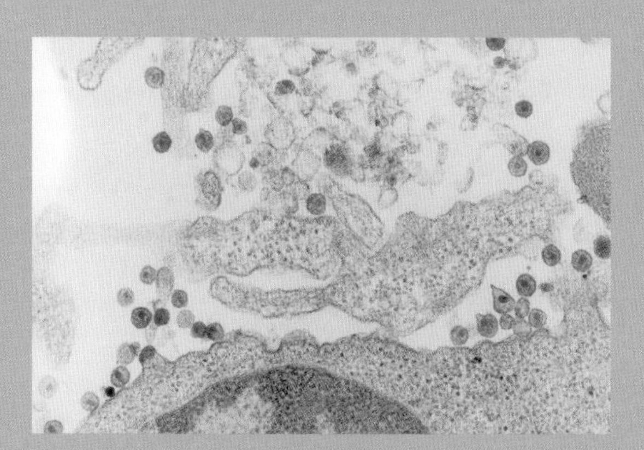

KEY DATES

1981: First cases reported
1983: Virus discovered at the Institut Pasteur
1985: Specific test developed
1985: Viral genome sequenced
1985: HIV-2 isolated
1996: First triple therapies introduced
2008: Nobel Prize in Physiology or Medicine awarded to Luc Montagnier and Françoise Barré-Sinoussi

Above: One of the very first photos of HIV-1 (the AIDS virus) taken on February 4, 1983. Partial section of a T lymphocyte infected by the virus isolated in a patient.

Effective treatment, but no cure

When infected with HIV, individuals can remain asymptomatic or suffer a primary infection. This initial phase of infection is characterized by non-specific, flu-like symptoms (high fever, muscle pain, headache, diarrhea), and is followed by an asymptomatic phase that sometimes lasts for several years. However, the virus is acting silently during this time, weakening the immune defenses and allowing opportunistic infections to take hold, caused by germs that would be harmless in healthy individuals. If left untreated, the disease develops into acquired immunodeficiency syndrome (AIDS), which is the final stage of HIV infection and is characterized by multiple bacterial, parasitic and fungal infections and susceptibility to certain cancers.

The therapy of choice combines several antiretroviral drugs—usually three, giving rise to the term "triple therapy". This therapy controls viral replication, so that the virus is no longer detected in the blood and the risk of transmission is therefore minute. It restores CD4 lymphocytes and more or less fully restores immune system efficacy. The sooner treatment is started the more effective it is, but it can never entirely eradicate the virus from the body; if treatment is stopped, viremia—or the presence of the virus in the bloodstream—follows within six to eight weeks. The persistence of HIV even during treatment is linked to a very low level of residual replication, in certain body tissues in particular, and to the survival and proliferation of latently infected reservoir cells that are beyond the reach of replication-blocking antiretroviral drugs.

THE ORIGINS OF AIDS: THE VIRUS, MONKEYS AND HUMANS

In 1984, barely three years into the pandemic, it was strongly suspected that the virus had originated in Africa. There were two reasons for this. The first epidemiological studies conducted in Zaire and Rwanda had revealed a high prevalence of HIV in these countries, and several hundreds of cases diagnosed in Europe appeared to have been contracted in Central Africa. Studies in the 1980s rapidly established that African monkeys had infected captive macaques. The belief that HIV was derived from a non-human primate virus began to emerge.

In Gabon in 1988, Martine Peeters and her team (French Research Institute for Development, Montpellier) identified two retroviruses in chimpanzees, which they called SIVcpz-GAB 1 and SIVcpz-GAB 2. These showed a marked genetic similarity to HIV-1. Ten years later, an extremely rare variant of the virus—HIV-1 Group N—was discovered in a 57 year-old patient in Cameroon by Institut Pasteur teams in Paris and Bangui. This variant was even more closely related to the simian pathogens SIVcpz-GAB 1 and 2, which infect a sub-species of chimpanzees, *Pan troglodytes troglodytes*, found in a forested region straddling Cameroon, Gabon, Equatorial Guinea and Congo.

We now know that HIV-1 is divided into four genetically distinct groups: M, N, O and P. Scientists generally agree that each of these groups was created when the disease passed from animals to humans, representing at least four occasions when the virus crossed the species barrier. There are seven such instances for HIV-2. Groups M and N are derived from viruses that infected *Pan troglodytes troglodytes* chimpanzees. More recently, in 2015, the team led by Martine Peeters showed that reservoirs of groups O and P have origins in gorillas living in South West Cameroon.

These studies are highly important for reconstructing the origins of the disease, understanding viral evolution and implementing an appropriate monitoring strategy.

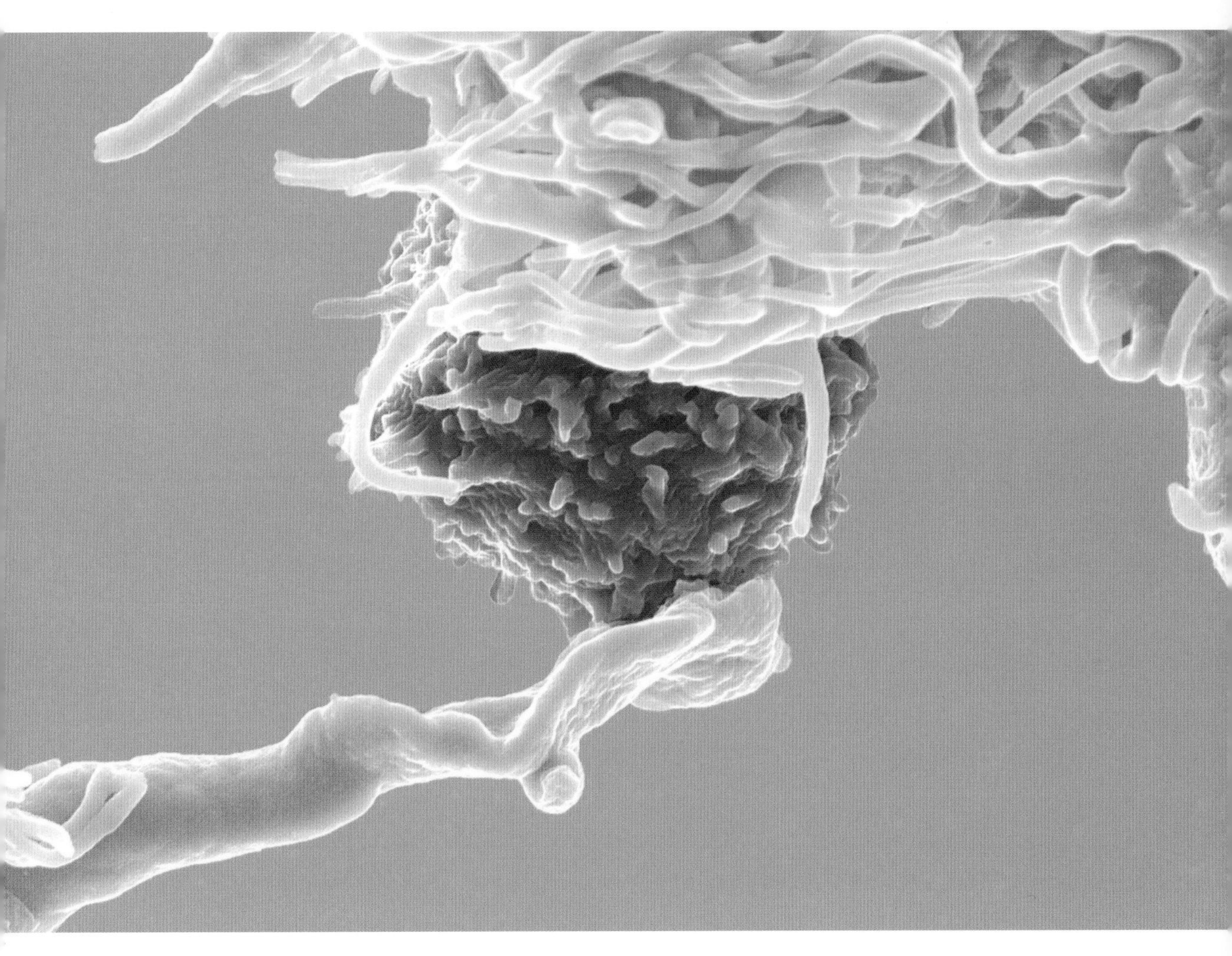

“Louis Pasteur understood that research aims not only to provide knowledge, but also tools and treatment for improving human life across the globe. With translational research the Institut Pasteur is following the path marked out by its founder.”

Françoise Barré-Sinoussi

Interaction between an HIV-infected T4 lymphocyte (in purple) and a dendritic cell (in green). These two immune system cells are targets for the virus—HIV exploits contacts between them to spread from one cell to the next.

FROM SEROLOGICAL DIAGNOSIS TO VIRAL LOAD

The first tests developed at the Institut Pasteur to diagnose HIV infection were serological tests. They consisted of detecting specific HIV antibodies in blood samples. Asymptomatic patients then became known as "HIV positive". These tests gave only an indirect picture of the presence of HIV, and provided no indication of its activity. Further, patients became HIV positive only after seroconversion—development of antibodies by the immune system—, which occurs between two weeks and three months after initial infection. This drawback also applies to the home-testing kits readily available in pharmacies, which, on the plus side, offer privacy as well as convenience as a single drop of blood or an oral swab is all that is required.
The invention of PCR (polymerase chain reaction), a genetic amplification technique that earned a Nobel Prize for its inventor, Kary Mullis in 1993, provided a means of measuring viral load, in other words establishing the quantity of virus in the blood. From a known sequence of some twenty complementary nucleotides on the DNA strand being investigated, PCR starts a chain reaction resulting in the production of millions of copies of the sequence under study. More specifically, RT-PCR is used for HIV since viral RNA first needs to be converted to DNA by reverse transcriptase.

Examining the results of a western blot showing the expression of an HIV-1 restriction factor in cells targeted by the virus, in different experimental conditions (2013).

In order to improve treatments against the backdrop of a pandemic that, although losing ground, has yet to be tamed, research on HIV remains a key priority for the Institut Pasteur and its international network. Research is transversal and multidisciplinary, and is carried out in partnership with a number of hospitals and national and international organizations, in particular the ANRS (French national agency for AIDS research). At the Institut Pasteur's Paris headquarters, this research involves a dozen teams from the departments of Virology, Immunology, and Infection and Epidemiology, with the Medical Center providing clinical research input.

> "At the Institut Pasteur, research on HIV ranges from the molecular and cellular level to the human body as a whole, and includes patient cohorts and populations of infected people as well as animal models."
>
> Olivier Schwartz, Head of the Virus and Immunity Unit at the Institut Pasteur

Tracking down the virus

During infection, HIV can transfer from an infected cell to a healthy cell. It takes advantage of the contacts established between lymphocytes, which it stabilizes to establish virological synapses. As well as providing an efficient transfer mechanism, these bridges allow HIV to evade antibodies. The team led by Olivier Schwartz, who manages the Virus and Immunity Unit, used fluorescent viruses to film cell-to-cell transfer in real time using video microscopy—a technique used on cell cultures or animal models to shed light on the dynamics of infection.

In a patient undergoing treatment, there is nothing to differentiate healthy CD4 lymphocytes from those harboring the virus. A decisive breakthrough was achieved in the identification of diseased cells by Monsef Benkirane and his colleagues at the Molecular Virology laboratory in Montpellier, as part of an ANRS project involving several teams including

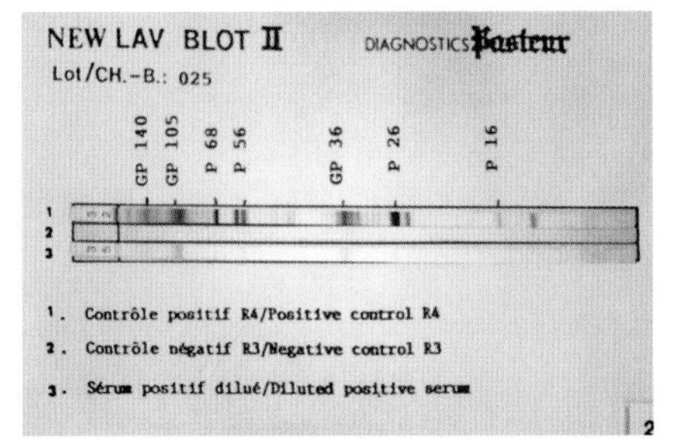

Western blot II assay for HIV diagnosis (1989).

ELECTRON MICROSCOPY

Viruses cannot be seen with normal optical microscopes due to their maximum magnification of x2,000. These microscopes produce an image by interaction with light, but image resolution is limited by the wavelengths in the light spectrum. Electron microscopy and the first experimental activities were inspired by quantum theory, and exploited the wave-particle duality of electrons. First used in the early 1930s, electron microscopy consists of replacing the light source with an electron beam whose wavelength is altered by accelerating voltages of between 20,000 and 300,000 volts. There are now two, complementary techniques: transmission microscopy, in which the beam sweeps over the specimen with up to 2 million fold magnification, and scanning microscopy, in which the electrons scan the specimen surface.
This technology was not initially considered suitable for living matter given the need for complete dehydration and almost total vacuum conditions before bombarding the specimen with electrons. In 1945, an electron microscope manufactured by French company *Compagnie* Générale de Télégraphie Sans Fil was installed at the Institut Pasteur. With this new tool, the team led by Pierre Lépine, with the help of a physicist, produced multiple images of bacteriophages and viruses, even overturning certain established classifications. Electron microscopy soon became an integral part of microbiology at the Institut Pasteur.

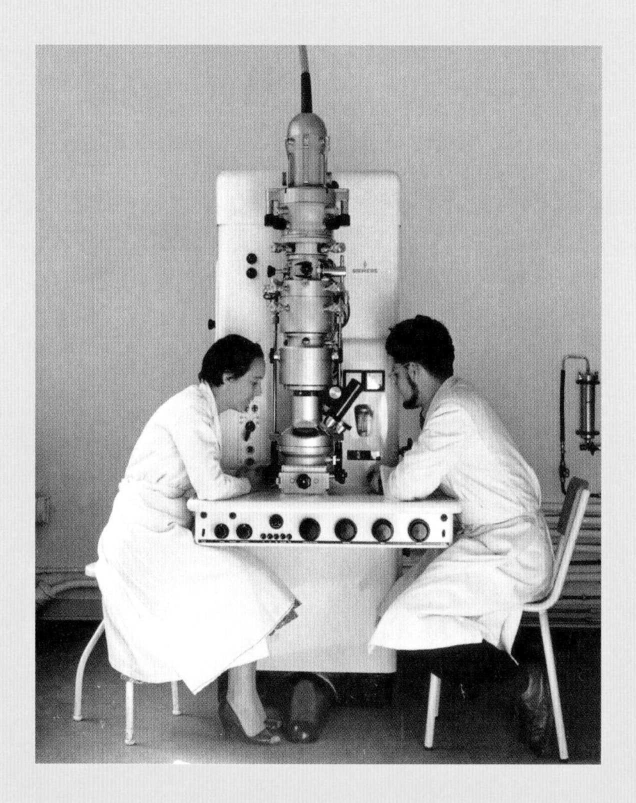

Odile Croissant and Charles Dauguet using a Siemens Elmiskop 1 electron microscope. Electron microscopy laboratory in the Virus Department directed by Pierre Lépine, Institut Pasteur, around 1960.

Olivier Schwartz's team at the Institut Pasteur. Using high throughput RNA sequencing (RNA-Seq), the scientists identified about a hundred genes expressed solely in infected cells, a dozen of which coding for surface proteins. Among these proteins, the CD32a molecule marks the majority of lymphocytes carrying a dormant, potentially reactivatable form of the virus. The discovery of a reservoir cell marker will enable the scientists to determine the location of the cells in the body, observe how they divide, and understand the viral latency mechanisms involved. In time, this could result in treatments aimed at eliminating these HIV sanctuaries.

HIV infection is also characterized by chronic inflammation and sustained activation of the immune system, which assist the virus by constantly producing the cells it targets—CD4 lymphocytes.

HIV controllers

Since the 1990s, scientists have focused on a small group of people infected with HIV who are naturally able to control HIV replication without treatment. The ANRS "HIV controllers" cohort is made up of HIV-infected people with undetectable viral loads. In these patients, the teams led by Olivier Schwartz and Hugo Mouquet, of the Humoral Response to Pathogens Five-year group at the Institut Pasteur, identified certain broadly neutralizing antibodies (bNAbs) able to block a number of HIV-1 strains and eliminate infected cells cultured in laboratory conditions. American teams have begun clinical trials using bNAbs. When injected into patients they significantly reduce viral load for about a month. The first clinical trials were conducted with a single antibody but, as with triple therapy, a combination of antibodies will probably be needed to avoid selection of mutant viruses. In addition to their neutralizing potential, these antibodies recognize GP120, the HIV envelope protein on the surface of reservoir cells, which they destroy via antibody-dependent cell-mediated cytotoxicity (ADCC). These antibodies constantly stimulate the immune system, producing vaccine-like effects.

ADCC (antibody-dependent cell-mediated cytotoxity): a reaction by which an immune system cell destroys a cellular target identified by the presence of antibodies. In the above case, neutralizing antibodies are directed against the GP120 HIV envelope protein exposed on the surface of reservoir cells, which are then recognized and destroyed.

Another cohort—the VISCONTI cohort—includes patients who were given early treatment and are still controlling viral replication more than ten years after treatment was stopped. At the Institut Pasteur, Asier Sáez-Cirión and his Viral Reservoirs and Control Group are working with these patients in an effort to understand what triggers viral latency and to develop therapeutic strategies to achieve remission.

Inasmuch as the reservoir cell becomes the therapeutic target, a parallel may be drawn with cancer cells, and this can lead to immunotherapies which have proven efficacious in cancer patients. Work is currently ongoing in this field.

Vaccine research and its hurdles

The Institut Pasteur works with a consortium of laboratories that are part of the Vaccine Research Institute (VRI), coordinated by Yves Lévy at Henri Mondor Hospital. Vaccine candidates are currently undergoing clinical trials. Previous attempts made in a number of countries had come up against two major hurdles: viral variability and the problems inherent in producing broadly neutralizing antibodies. Françoise Barré-Sinoussi explained that these challenges, already known for some years, are associated with gaps in the scientific community's understanding of vaccination principles in general (see page 58). "We have not fully understood all the immunological mechanisms required to confer protection." Hence the importance of fundamental immunology research at the Institut Pasteur and in many laboratories worldwide.

Another strategy consists of using antibodies to target antigen-presenting dendritic cells. By eliminating cells that contain fragments of the virus and activate naive T lymphocytes, we hope to reduce production of the CD4 lymphocytes targeted by the virus, thereby preventing an excessive immune response.

Antigen: a substance the body identifies as foreign (a bacterium, virus, etc.) that may trigger an immune response and the formation of antibodies to eliminate it.

Dendritic cell: a large, specialized cell that digests microbes and foreign substances and presents resulting fragments to other immune system cells.

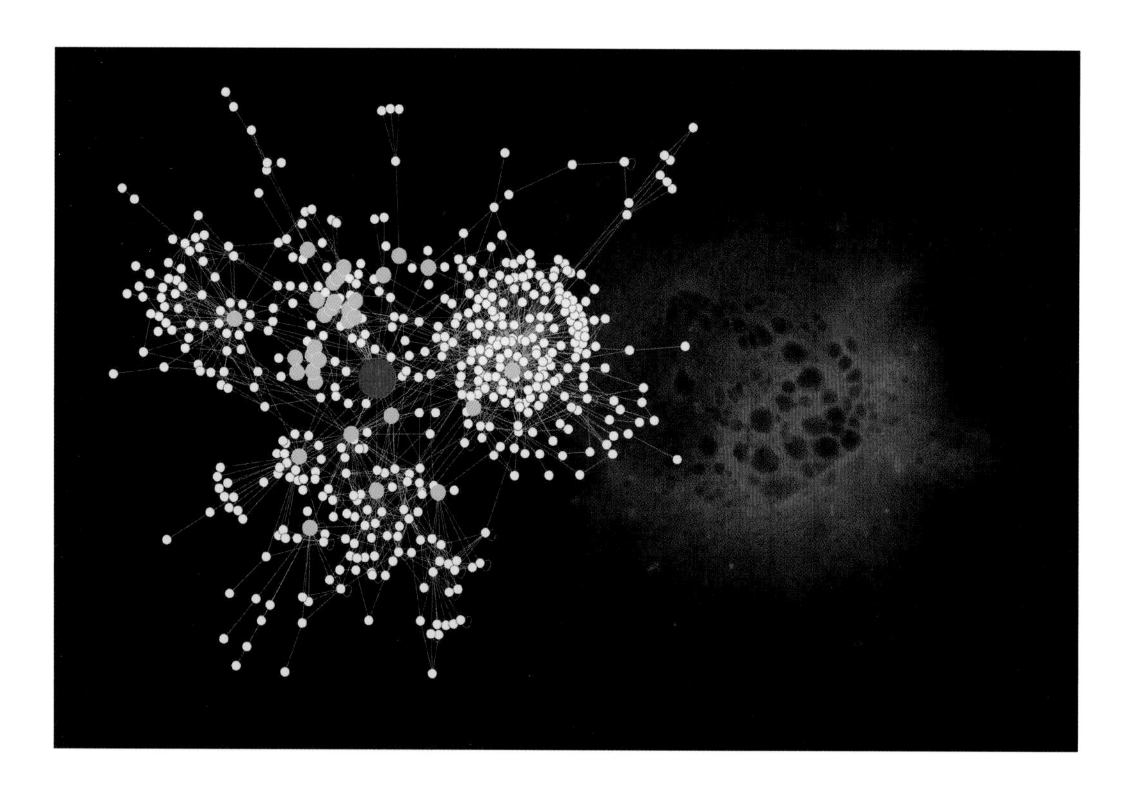

Vaccine research: using the measles vaccine to tackle other diseases. The measles virus multiplies by establishing a network of interactions with proteins in our cells. By redirecting these connections, the aggressive action of the virus can be reprogrammed to target other diseases.

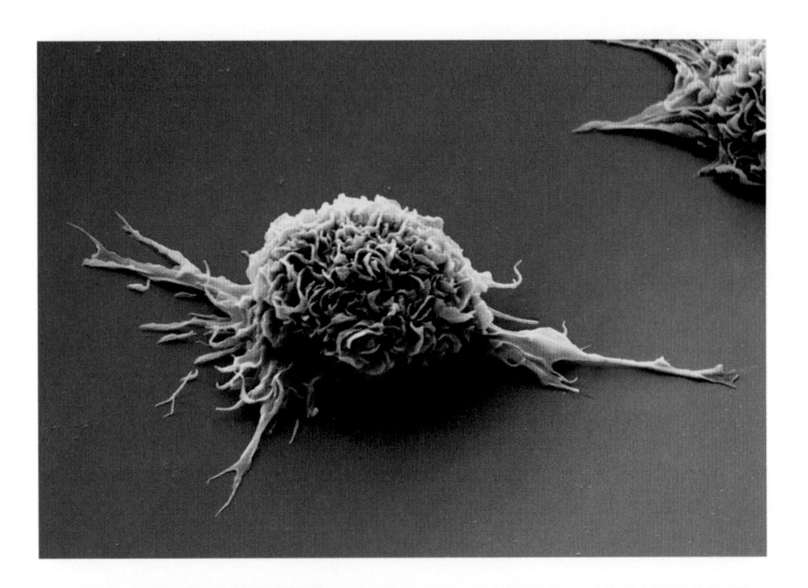

THE IMMUNE SYSTEM: A DELICATE BALANCE

The immune system is made up of all the organs, tissues and cells that protect us from environmental invaders—most importantly invasion by pathogens. But that's not the whole story. It also plays a part in homeostasis, for example by getting rid of dead cells or acting early to prevent the spread of cancer cells. Paradoxically, this protection system tolerates some foreign invaders, such as the billions of bacteria, viruses and fungi that inhabit our digestive system, or a growing fetus, which shares only half its genes with the pregnant mother. However crucial its activation in the event of invasion by unknown microorganisms, unchecked it may give rise to allergies or autoimmune diseases, which are on the rise in all developed countries. In a healthy person, this highly dynamic system comprising cells that move, interact, exchange information and produce a number of inflammation mediators is held in a delicate balance between stimulation and regulation, involvement and detachment.
A distinction is made between innate and adaptive immunity. Innate immunity is characterized by a series of responses designed to eliminate outside invaders as quickly as possible. It reacts to microbial components that have remained intact throughout evolution, and depends on cells with the ability to "engulf" the invader (macrophages, dendritic cells, polymorphonuclear neutrophils) or kill it (NK, or natural killer cells). Adaptive immunity is a more specific response that "adapts" to rapid changes in the invader. It depends on B lymphocytes, which produce antibodies, and T lymphocytes, which can trigger destruction of infected cells or cancer cells. This is a second line of defense, requiring cell selection and maturation. It also "remembers" the invader.
This is, of course, merely the broadest outline of the mechanism. More than a century after the first discoveries by Ilya Mechnikov (see page 14), immunology has become a complex and transversal science, and is a highly active field of research. At the Institut Pasteur in Paris, the Immunology Department comprises 14 research units, 2 technical platforms and 166 scientists, under the direction of Gérard Eberl. In addition to the work conducted there and the innovative technologies used—such as imaging procedures for dynamic filming of immune cell movement and interaction—, the training of a new generation of immunologists is a primary goal.

Homeostasis: the biological process by which conditions (temperature, blood sugar level, etc.) are controlled to maintain levels that are beneficial for the body.

A dendritic cell, which plays an important role in the immune system, viewed using scanning electron microscopy.

Anticipating and Controlling

Emerging Infectious Diseases

EMERGING DISEASES
IN FIGURES

11,323 DEATHS
from Ebola virus disease out of a total 28,000 notified cases in Guinea, Liberia and Sierra Leone in 2014-2016

MORE THAN 100
articles published about the Zika virus between 2014 and 2017 in the Institut Pasteur International Network

ALMOST 40 MILLION
deaths from HIV; the AIDS epidemic is described by WHO, in 2017, as the world's most significant public health challenge

In the wake of the Second World War, medical advances and the triumph of penicillin fueled hopes for an aseptic future world devoid of infection. The warnings of Institut Pasteur scientist Charles Nicolle—winner of the Nobel Prize for Medicine in 1928 for identifying the louse as the vector of typhus—had all too soon been forgotten. Nicolle, who was Director of the Institut Pasteur in Tunis and had clearly been influenced by Darwinism during his spells in Mechnikov's laboratory, believed that microbes could evolve. As the author of several works on the history of epidemics, he had learned that pathogens have a "mosaic of powers", and are able to undergo change via rearrangement and as a result of chance environmental conditions. Charles Nicolle was convinced that infectious diseases "are born, live and die", and that new, previously unknown epidemics would emerge. "When new outbreaks emerge, we will be aware of their existence only when the number of reported cases forces us to acknowledge the presence of a new evil; the circumstances surrounding their emergence will be as elusive to us as if we were researching the origins of a thousand year-old disease." Due to his powerful insight, combining microbiology and a historical perspective, Charles Nicolle is regarded as one of the very first thinkers in the field of disease emergence.

Opposite: Hélène Sparrow (1891-1970) and Charles Nicolle (1866-1936) on a field trip to investigate epidemic typhus in Mexico (1931).

Charles Nicolle in his laboratory at the Institut Pasteur in Tunis.

In the early 1980s, the AIDS pandemic permanently laid to rest the illusion that pathogens had been conquered. What was this new disease? What was causing it? Where did it come from? The questions being asked at that time were the same as those asked throughout history about countless epidemics—by the Greeks when an unknown disease plunged Athens into mourning in the time of Pericles, by the Venetians when the Black Death ravaged their city in 1348, and by the Parisians when cholera swept through Europe in the 19th century. For a rational answer to these questions, the imprecise notion of a "new" disease had to be replaced by the notion of an "emerging" disease. Arnaud Fontanet, who heads up the Epidemiology of Emerging Diseases Unit at the Institut Pasteur, defines an emerging disease as follows: "The appearance, within a human population, of a new pathogenic agent derived from an animal or environmental reservoir or following the genetic modification of an existing pathogen". Such a disease can also experience a shift in geographical location. This vision, which has its parallels with the writings of Charles Nicolle, may apply to a disease that has already, at a previous point in history, made its mark on a given region. This is known as a re-emerging disease. It can therefore also potentially apply to non-communicable diseases. This chapter, however, will concentrate on emerging infectious diseases and their two triggers—on the one hand the nature and genealogy of pathogens, and on the other the history of outbreaks, and their origins and proliferation factors. Charles Nicolle was absolutely right: gaining control over an outbreak demands an understanding of how it started.

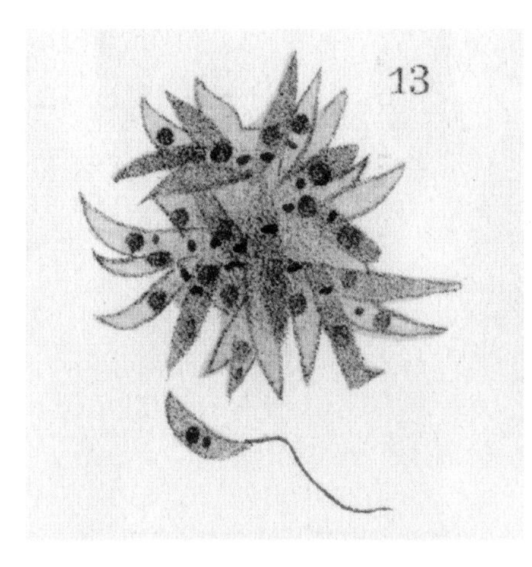

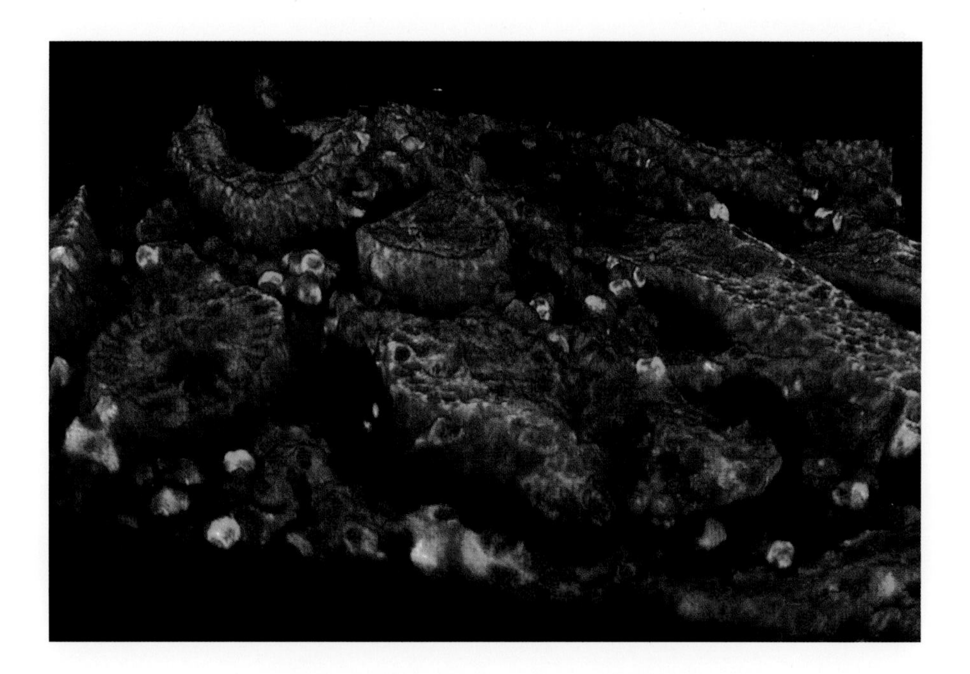

Top image: Description of "Leishman bodies" by Charles Nicolle. Drawing taken from a plate from the archives of the Institut Pasteur in Tunis, February 1908.

Bottom image: Intestinal cells producing a cytokine that confers protection against AIDS in African green monkeys infected by the Simian immunodeficiency virus (SIV).

OUTBREAKS AND PASTEURIANS

The Institut Pasteur International Network operates on every continent and is widely experienced in all aspects of epidemiological surveillance and analysis of pathogenic strains. It is on permanent standby and can be mobilized quickly to respond to an outbreak wherever it may arise. Pasteurians today are carrying on the mission started by their elders in conditions that were far more hazardous.

Take, for example Alexandre Yersin, who arrived in Hong Kong on June 15, 1894, armed simply with a microscope and small autoclave. In the company of two hastily recruited helpers, the microbiologist combed the deserted city in search of corpses for the samples he needed. He was denied permission to examine the dead by the local authorities whose preference went to a Japanese team that had arrived a few days earlier. Yersin was determined not to give up. Within two days he had built a hut to house his laboratory and set to work. "A few, carefully distributed piasters" opened doors and gave him access to the body of an unlucky individual who had died some hours earlier. A sample from an inguinal bubo, quickly prepared and placed under his microscope, revealed "a veritable soup of microbes". There, beneath the eyepiece, was Yersin's bacillus. He had just discovered the culprit responsible for one of the greatest evils of all time, and raised the curtain on the plague that swept through Europe and remains etched in the collective memory to this day. In the wake of this discovery, he showed that the epidemic was spread by rats, but failed to discover the 'missing link' between rodents and humans, namely fleas.

We should also spare a thought for Émile Marchoux, who was sent to West Africa in 1895 by Émile Roux to set up the very first African microbiology laboratory in Saint-Louis, Senegal. The former army doctor used Pasteurian methods to study tropical diseases, many of unknown cause. There was no shortage of endemic diseases and epidemics in that particular region, including malaria, yellow fever, sleeping sickness, leprosy, amoebic dysentery, to name a few. Despite rudimentary means and having no fixed base, Marchoux laid the groundwork, publishing in tropical medicine journals while providing care to the local people.

The examples of these two pioneers illustrate the impressive work accomplished in extremely difficult conditions. Over time, many other Pasteurians—both famous and unknown—brought their expertise and know-how to the front line, where the Institut Pasteur is today, striving daily to further knowledge and aid populations.

Top: Hong Kong—Dr. Yersin at the entrance to his thatched hut (1894).
Opposite: Yellow fever virus (17D vaccine strain) in human cells (1999).

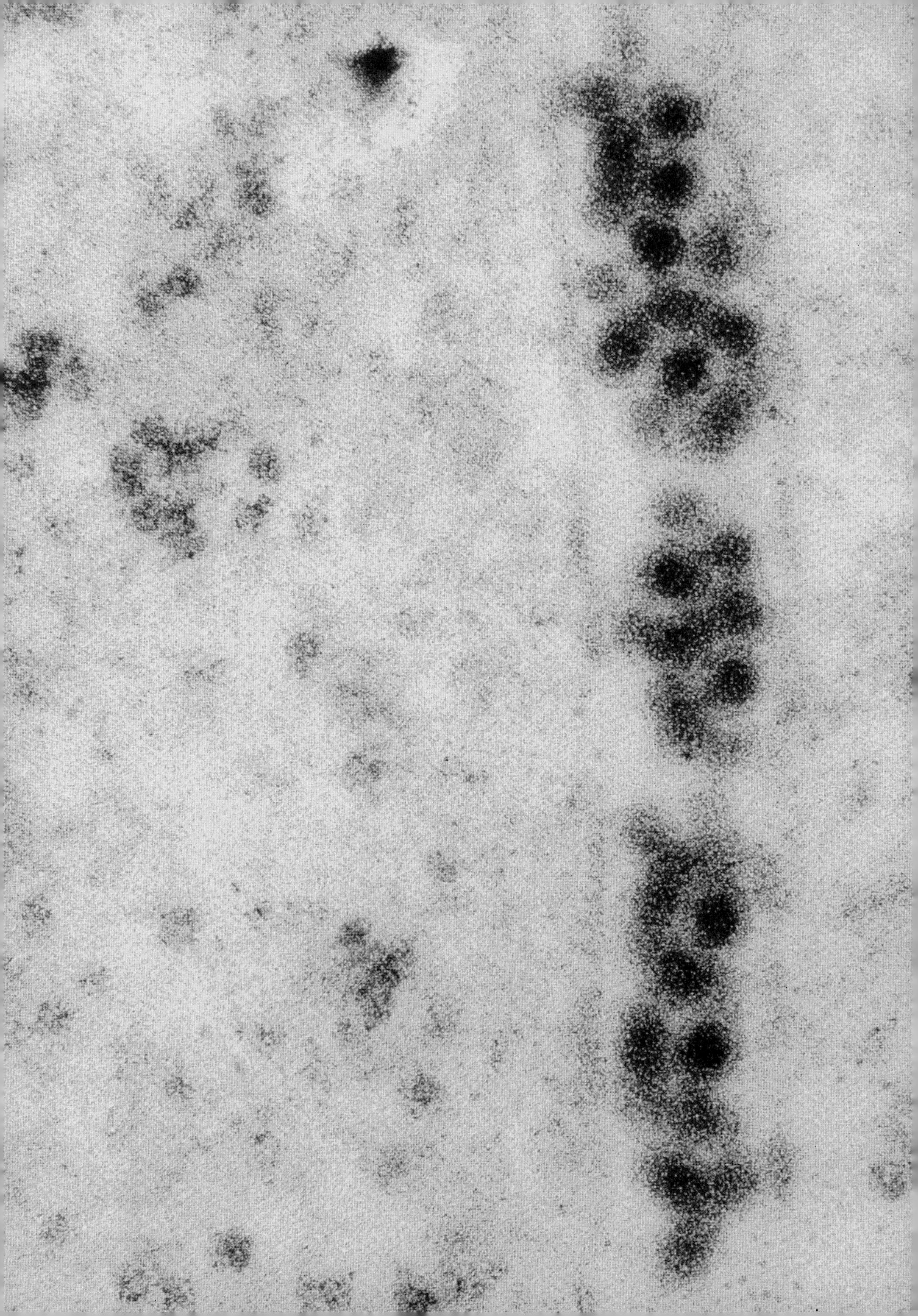

Ebola—an often fatal hemorrhagic fever

First seen in 1976, when an outbreak occurred in the village of Yambuku, near the Ebola River in the Democratic Republic of the Congo, this virus belongs to the filovirus family. Ebola's high death rate quickly became apparent in this first outbreak, with 280 deaths (88% of infected people) in just a few weeks. At the same time, some 2,000 kilometers away, a different strain struck South Sudan. Successive outbreaks in the forested regions of Central Africa led to the identification of five different species, which were named after the geographical regions in which they were isolated: Zaire, Sudan, Reston (USA), Taï Forest (Côte d'Ivoire), and Bundibugyo (Uganda). Scientists have shown that fruit bats are probably the natural hosts of these viruses, which are transmitted to humans after infecting intermediate hosts—usually game hunted for meat and, in particular, monkeys susceptible to the infection.

The incubation period for Ebola is 2 to 21 days after exposure. The disease is characterized by acute, non-specific, flu-like symptoms: severe fever, muscle pain, headache and sore throat. This is followed by diarrhea, vomiting, impaired kidney and liver function and, in some cases, internal and external bleeding predictive of poor prognosis. Depending on the strain and the outbreak in question, between 30% and 90% of patients die from acute infection. Diagnosis is confirmed by laboratory tests, in particular the RT-PCR technique, carried out on high-risk biological samples that must be handled according to strict safety procedures.

Until the 2014 outbreak, there was no specific vaccine or treatment for Ebola. Care focused on symptom relief, in particular rehydration of patients prone to excessive fluid loss due to vomiting and diarrhea.

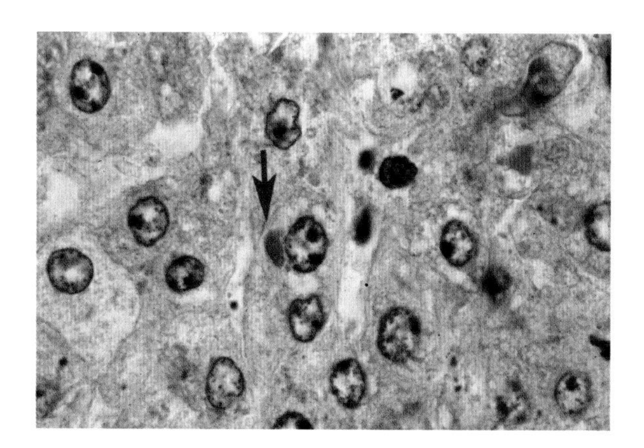

Above: The Ebola virus in the cytoplasm of liver cells (2000).

Opposite: Prof. Pierre Sureau from the Institut Pasteur examining a patient during the 1976 Ebola fever outbreak in Kinshasa.

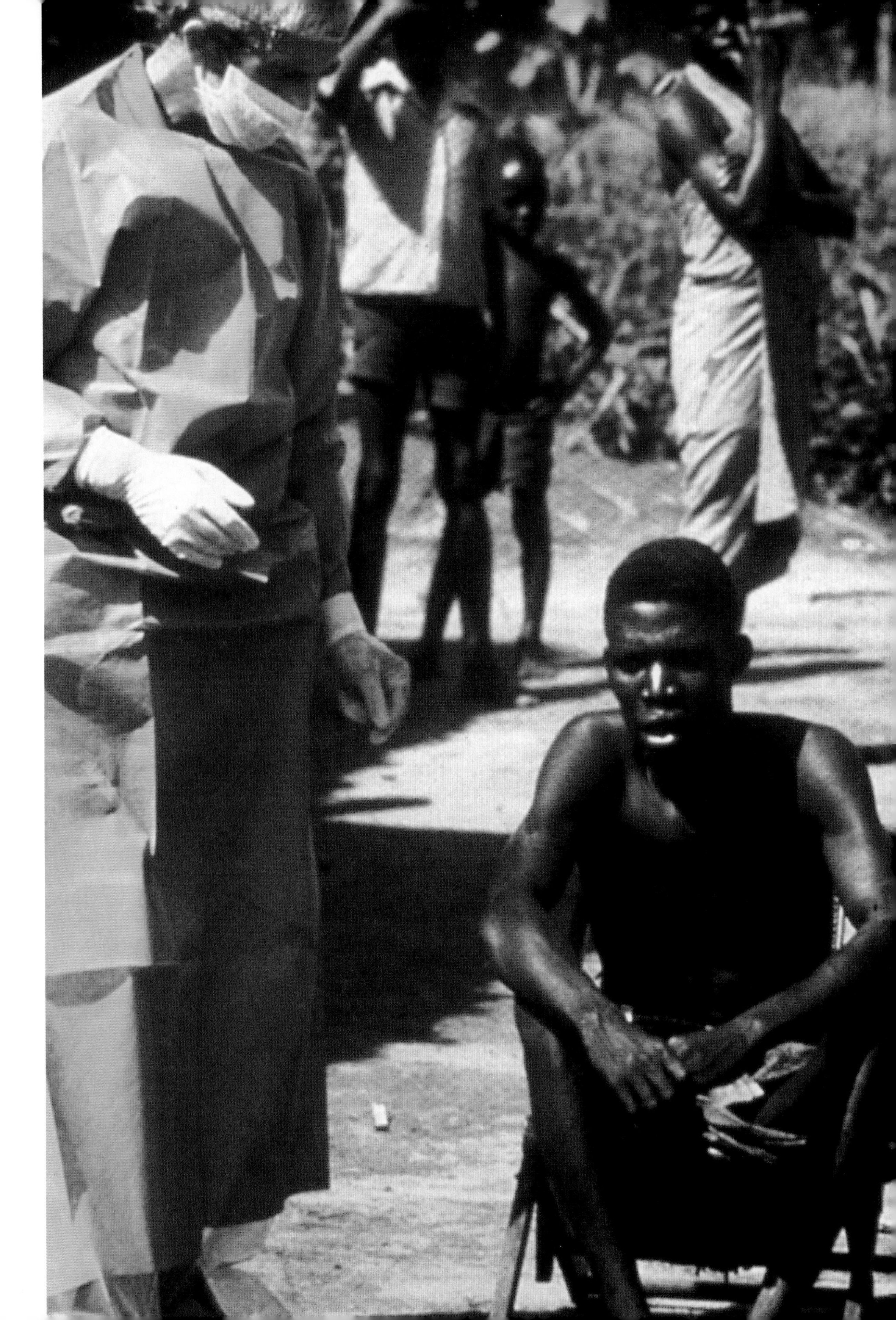

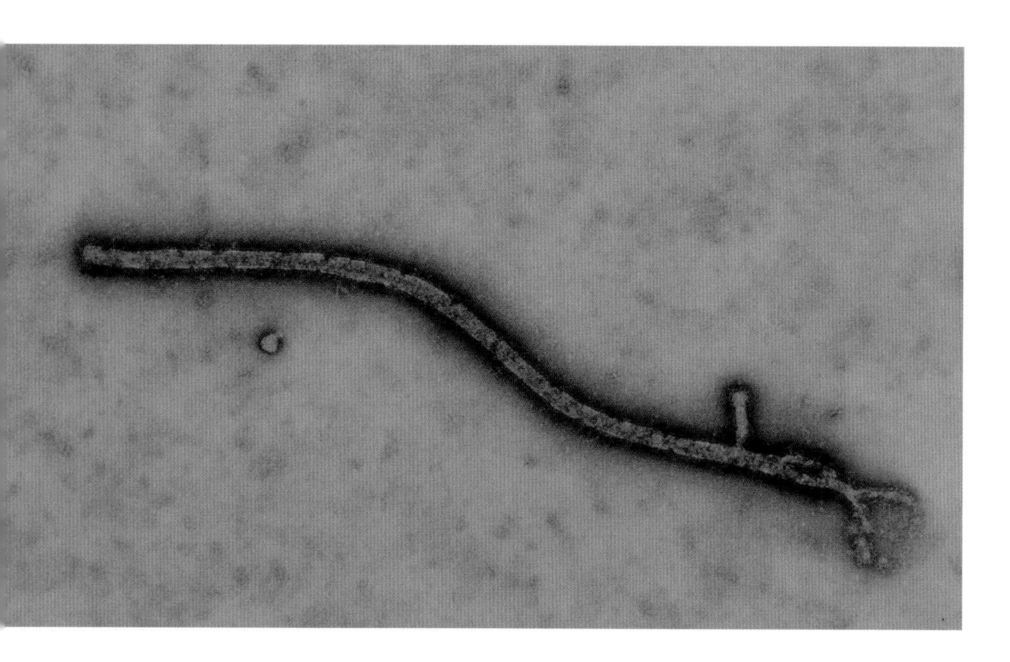

The Institut Pasteur at the forefront of the Guinea outbreak

In March 2014, the first blood samples from Guinea arrived at the National Reference Center (CNR) for Hemorrhagic Fevers led by Sylvain Baize in Lyon, which operates a BSL-4 high containment facility. The outbreak had already been brewing in the region for several months. Later, it became known that it had started in the village of Meliandou, taking the life of a 2 year-old boy in late December 2013. In this poor and remote forest region straddling Liberia and Sierra Leone, ravaged by years of civil war, the first cases went unnoticed—overlooked in a country plagued by countless deadly diseases where Ebola had yet to be seen.

On March 21, the CNR identified a filovirus thought to be either Ebola or the Marburg virus. The following day, the Institut Pasteur Laboratory for Urgent Response to Biological Threats (CIBU), led by Jean-Claude Manuguerra, confirmed the pathogen to be a strain of Zaire ebolavirus (ZEBOV)—the most virulent of them all. On March 23, WHO raised the alert.

The Institut Pasteur immediately mobilized its international network, particularly in Senegal and France. Amadou Sall's team at the Institut Pasteur

The Ebola virus viewed using transmission electron microscopy.

in Dakar was commissioned by WHO to deploy the first high-tech laboratory equipped for biological diagnosis of the disease at Donka Hospital in Conakry. In November 2014, in association with Inserm and the French Research Institute for Development (IRD), Sylvain Baize's team installed a second facility—a 60-bed treatment center in Macenta, in Guinea's forest region, run by the Red Cross. The CIBU team tested thousands of samples, and trained Guinean technicians in sampling and biological diagnostic methods. Jean-Claude Manuguerra's team, already involved in a program co-funded by MSF for the rapid detection of bacteremia in young children in Sub-Saharan Africa, was successful in developing a rapid diagnostic test using LAMP (loop-mediated isothermal amplification) technology. The test kit was cheaper, lightweight and, above all, produced results within just 15 minutes.

The Institut Pasteur reinforced its commitment with the establishment of an Institut Pasteur in Conakry. The decision was made in response to the Ebola crisis in 2014, in efforts to tackle emerging outbreaks on a long-term basis. The first stone of the Institut Pasteur in Guinea—the 33rd member of the Institut Pasteur International Network—was laid on November 11, 2016. The institute is directed by Noël Tordo.

CIBU: MICROBIOLOGICAL EMERGENCIES

The Laboratory for Urgent Response to Biological Threats (CIBU), led by Jean-Claude Manuguerra, was set up in 2002 by the French General Directorate of Health (DGS) and the Institut Pasteur to respond primarily to bioterrorist threats. Although the CIBU has fortunately never been called upon to handle malevolent acts, it has proved efficient in the field of outbreak emergencies. Since 2008, its laboratories have come under three divisions: virology, bacteriology and entomology. The CIBU can be called upon 24/7, and its mission is to identify and characterize the pathogen responsible, sequence its genome and develop diagnostic tests (see page 176).

Above left: Medical staff prepare to go to the isolation ward during the Ebola virus outbreak in Mekambo, Gabon, in 2001.

Above right: Building housing the high-technology diagnostic laboratory set up by the Institut Pasteur at the Macenta treatment center in Guinea's forest region during the Ebola crisis in November 2014.

> "Nobody can predict where or when the next epidemic will strike. In the face of this threat, the Outbreak Investigation Task Force and the Institut Pasteur International Network are particularly well prepared for a rapid and effective response and the immediate development of a research pipeline."

Arnaud Fontanet, Head of the Epidemiology of Emerging Diseases Unit and Director of the Center for Global Health at the Institut Pasteur

The causes of a health crisis

Ebola took a heavy toll. In June 2016, when WHO announced the official end to the Ebola outbreak, Guinea, Liberia and Sierra Leone had been badly affected, with at least 28,000 cases reported and 11,300 deaths. Although Ebola had occurred several times since 1976, it had never had such catastrophic consequences. There are many reasons for this grim statistic, including poverty in a forest region suffering ecological imbalance due to years of conflict, disruption within the health service, emergence of the disease at the crossroads of international trade routes connecting the capitals of three adjoining countries, the virulence of ZEBOV within a highly mobile population unaware of the disease, and lastly the difficulties inherent in winning the trust of the populations affected.

The virus is transmitted from human to human through contact with blood and bodily fluids. It is most contagious when the symptoms appear, but viral load is still high at the time of death, and the body remains infectious

after death. Funeral rites, which involve washing the body, represent a significant contamination risk. With the help of ethnologists, health workers need to inform populations and win their trust in order to ensure that the requisite hygiene and prevention measures are followed. This task is far from easy during an epidemic, when rumors tend to run rife and outside intervention appears as a threat to ancestral traditions. It requires an ability to reassure and adopt a more caring attitude, allow families to mourn for their loss, cooperate with local authorities and adopt a flexible approach when establishing protocols, while at the same time enforcing stringent measures to stop transmission. At the Institut Pasteur, the team led by Tamara Giles-Vernick, who heads up the Medical Anthropology and Environment group, is studying animal-to-human transmission and human-to-human transmission through the use of questionnaires and qualitative interviewing methods, to supplement the quantitative epidemiological data available. To understand the dynamics of Ebola, the Institut Pasteur in Dakar and the Mathematical Modeling of Infectious Diseases Unit in Paris, led by Simon Cauchemez,

Examining a bat in a BSL-3 laboratory at the Institut Pasteur in Bangui in 2012. Fruit bats are thought to be the natural hosts of the Ebola virus.

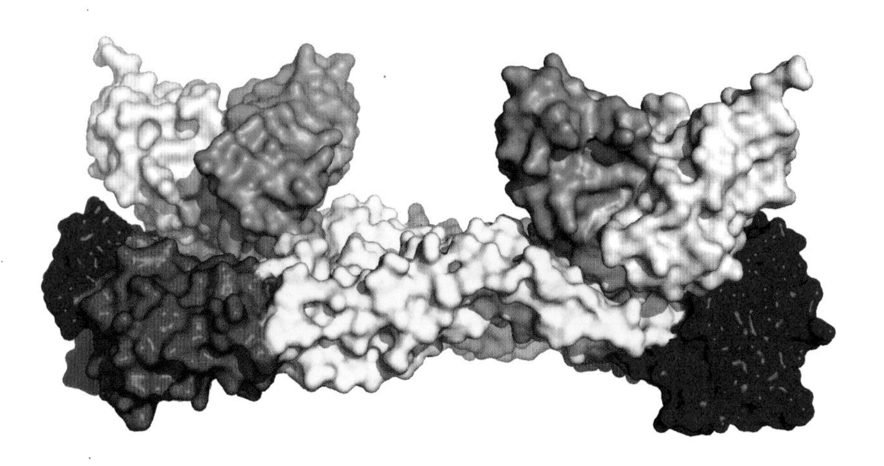

ZIKA: DECEPTIVELY BENIGN

Zika virus (ZIKV) is an RNA flavivirus. This family comprises a number of pathogens, including the causative agents of dengue, yellow fever and Japanese encephalitis. Spread by mosquito bites, it was first identified in 1947 in macaques, and subsequently in humans in 1954. The virus was known to be widespread in Asia and Africa, but in 2007 it hit Yap Island in Micronesia, where 70% of the population contracted a seemingly benign disease initially thought to be of little consequence.
In 2013, a dengue-like outbreak occurred in French Polynesia. This outbreak, which affected half the islands' population, was actually due to an Asian strain of ZIKV identified by the Institut Louis-Malardé in Papeete. It was characterized by neurological complications and Guillain-Barré syndrome. The ensuing paralysis, which is ascending and always regressive, requires intensive care admission in 40% of cases due to transient respiratory muscle involvement. The Epidemiology of Emerging Diseases Unit, led by Arnaud Fontanet, was operational in the region from January 2014 onwards and, in collaboration with the Institut Louis-Malardé, French Polynesia Hospital and the Public Health Surveillance Office, established a link between the virus and the 42 reported cases of Guillain-Barré syndrome.
In November 2015, when the epidemic hit Brazil with unexpected intensity, local health authorities were alerted to cases of abnormal brain development in newborns. A review of the records by the Polynesian teams, in collaboration with the Institut Pasteur teams of Arnaud Fontanet and Simon Cauchemez, showed that the malformations were also seen in the Polynesian archipelago. The scientists estimated a 1% risk of microcephaly if infection occurred during the first trimester of pregnancy. With its limited, well-documented population, efficient health care network, and cases clustered in space and time, Polynesia became a "Zika laboratory", providing WHO with rapid access to information on ZIKV pathogenicity in 2016.
While still raging in Polynesia the virus hit New Caledonia, where the Research and Expertise Unit on Dengue and other Arboviruses, led by Myrielle Dupont-Rouzeyrol, revealed that the virus remains for a longer time in urine than in the blood, thereby facilitating diagnosis.
When Zika arrived in Latin America, the Institut Pasteur in French Guiana sequenced the strain in circulation and identified it as the Polynesian strain. The Institut Pasteur in Dakar, which was already researching the circulation of ZIKV between mosquitoes and primates, is now working with other international network members, in particular the team led by Anna-Bella Failloux in Paris, on a project to assess the transmission capability of the different mosquitoes, and is studying the diversity of strains circulating on the various continents. It would be impossible to mention all the projects relating to ZIKV, but worthy of note is the work of Félix Rey in the Structural Virology Laboratory in collaboration with the CNRS and Imperial College, which has shown a neutralizing antibody to be effective against both Zika and dengue viruses.
With over 100 publications since 2014 by 16 of its members, the activity surrounding the Zika outbreak is further testimony to the drive and efficiency of the Institut Pasteur International Network.

3D structure of the Zika virus envelope protein (red, yellow and blue) in complex with a neutralizing antibody (in green and white).

reconstructed the Ebola transmission chain in the urban population of Conakry between February and August 2014. Transmission routes changed during the course of the epidemic: in March, 15% of infections were due to funeral rites and 35% occurred in hospital; these figures fell to 4% and 9% respectively from April onwards with the introduction of safer funeral practices and the opening of a treatment center.

Two other recently identified factors also favored the spread of the virus. During the outbreak, which was the longest ever known, scientists from the University of Nottingham, and Etienne Simon-Lorière from the Functional Genetics of Infectious Diseases Unit at the Institut Pasteur, analyzed 1,600 sequences of the Ebola virus from samples taken in Guinea, Sierra Leone, Liberia and Mali. They identified a strain with a greater ability to infect human cells than the cells of bats—the natural reservoir of the disease. This adaptation, which occurred early on in the outbreak, was thought to originate from a single mutation, known as A82V, which altered the viral envelope protein.

The second factor came to light during survivor follow-up. Whereas doctors no longer considered Ebola survivors to be contagious, scientists at the IRD, Inserm, the Institut Pasteur, and their Guinean partners showed that ZEBOV can persist in sanctuary sites, such as eyes, breast milk and, in particular, testicles. The disease is therefore liable to spread through sexual activity for weeks or even months following recovery.

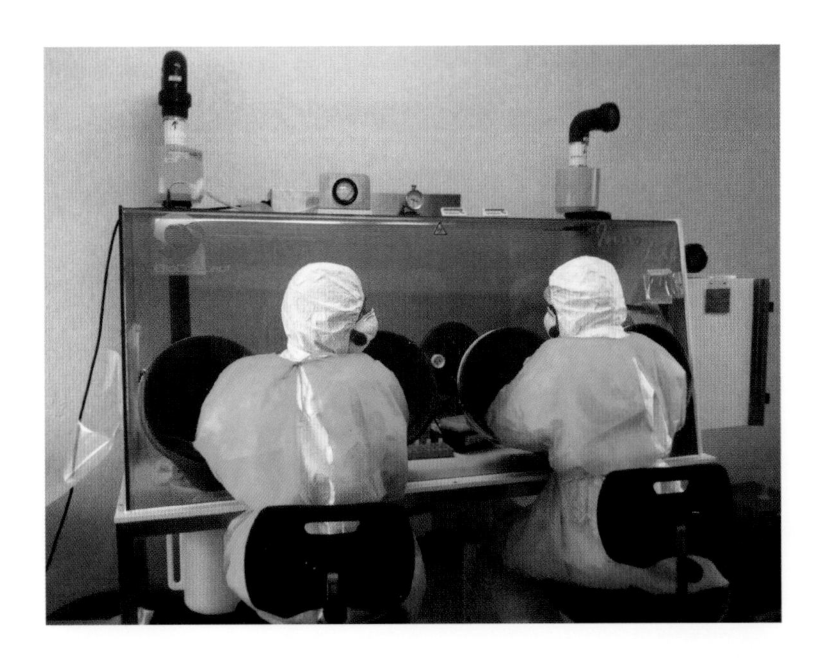

Ebola crisis: the high-technology diagnostic laboratory set up by the Institut Pasteur at the Macenta Ebola treatment center in Guinea's forest region in November 2014. The first team of volunteers sent by the Institut Pasteur to set up this laboratory was composed of Sylvain Baize, Head of the Biology of Viral Emerging Infections Unit, and two of the unit's senior technicians.

Vaccine hope

To monitor the pathogen, a consortium of teams coordinated by the Institut Pasteur's Outbreak Investigation Task Force stepped up efforts to sequence the strains isolated in Guinea and monitor their development. A genetically-engineered recombinant vaccine, called rVSV-Zebov, was tested on 11,841 people in Guinea. The vaccine was developed by a Canadian team, and would appear to be safe and provide early protection. However, it is effective only against Zaire strains and the duration of immunity is not yet known. At the Institut Pasteur, a preventive vaccine is also being developed by the Viral Genomics and Vaccination Unit, led by Frédéric Tangy, in collaboration with the Biology of Emerging Viral Infections Unit, led by Sylvain Baize.

In addition to the preventive vaccine, the team led by Pierre Charneau's Molecular Virology and Vaccinology Unit is working on a therapeutic vaccine strategy.

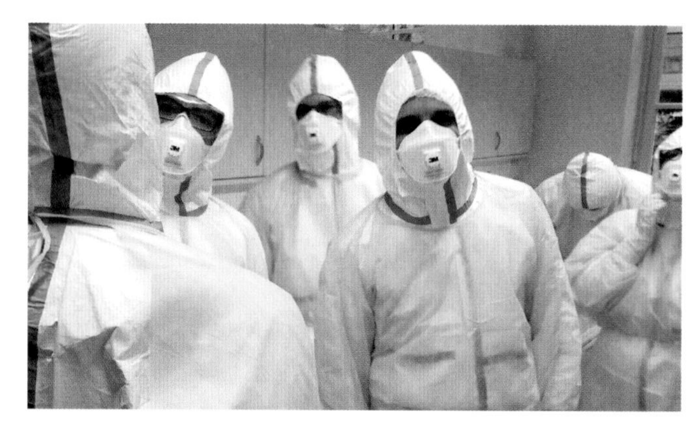

A TASK FORCE TO RESPOND TO EMERGING DISEASES

The Outbreak Investigation Task Force (OITF) was developed within the Center for Global Health led by Arnaud Fontanet in 2015 to provide a response to epidemic emergencies. The task force is coordinated by epidemiologist Maria Van Kerkhove. It is supported by the Institut Pasteur International Network, which is involved in surveillance and response to emerging diseases through multidisciplinary expertise provided by clinicians, epidemiologists, virologists, entomologists, and veterinarians, etc. The teams under the task force share their experience, knowledge and specific expertise, and are ready to respond rapidly and collaboratively in the event of a crisis. Depending on particular local issues, the OITF also ensures that effective, appropriate tools are available to all involved. In the event of an outbreak, its role is to assist local authorities and coordinate the collection and analysis of microbiological and epidemiological data. As well as providing an immediate response, it undertakes research on diagnostic tests, treatments, vaccines and epidemiological studies.
The task force was called upon to handle outbreaks of Ebola in Sierra Leone, Zika in Polynesia, avian influenza in Cameroon, and MERS-CoV (Middle East respiratory syndrome coronavirus) in Saudi Arabia and South Korea.

Ebola crisis, 2014. Training up volunteer scientists at the Institut Pasteur in Paris in November 2014 before they head out to work in the high-technology diagnostic laboratory set up by the Institut Pasteur at the Macenta Ebola treatment center in Guinea's forest region.

A SERIES OF EMERGING DISEASES IN MANY SHAPES AND FORMS

Although they do not always make the headlines, emerging infectious diseases are a regular occurrence and come in various shapes, sizes and severity. It is estimated that 70 to 75% of them are zoonotic—in other words transmitted from animals to humans. The table below illustrates the rich diversity of the key diseases.

DATE	DISEASE	CAUSE	LOCATION	RESERVOIR	VECTOR
1965	Japanese encephalitis	Japanese encephalitis virus (flavivirus)	Japan	Pigs and birds	*Culex* mosquitoes
1969	Lassa fever	Lassa virus (arenavirus)	West Africa	Multimammate rats	
1976	Ebola hemorrhagic fever	Ebola virus	Yambuku (Congo)	Bats Intermediate hosts: monkeys	
1976	Lyme disease	*Borrelia burgdorferi* bacterium	Old Lyme (Connecticut, USA)	Rodents, deer, squirrels	Ticks
1976	Legionnaire's disease	*Legionella pneumophila* bacterium	Philadelphia (USA)	Domestic hot water systems, cooling towers, natural stagnant water	
1983	HIV/AIDS	Human immunodeficiency virus (HIV)	San Francisco (first identified cases)	Chimpanzees, gorillas	
1986	Bovine spongiform encephalitis, or mad cow disease	Prion	United Kingdom		Bone meal
1999	Acute respiratory syndrome or encephalitis	Nipah virus (henipavirus)	Kampung Sungai Nipah (Malaysia)	Bats Potential intermediate hosts: pigs	
2002	Severe acute respiratory syndrome (SARS)	SARS-CoV (coronavirus)	Canton (China)	Bats Intermediate hosts: civets	
2005	Flu-like symptoms, joint pain, rare neurological disorders	Chikungunya virus	Reunion Island (Indian Ocean)		*Aedes albopictus* and *aegypti* mosquitoes
2009	Influenza	H1N1 virus	Mexico	Birds Intermediate hosts: pigs	
2012	Respiratory syndrome	MERS-CoV (coronavirus)	Saudi Arabia	Bats Intermediate hosts: dromedaries	
2013	Ebola hemorrhagic fever	Ebola virus	West Africa	Bats Intermediate hosts: monkeys	
2016	Flu-like symptoms, Guillain-Barré syndrome, microcephaly	Zika virus (flavivirus)	Brazil		*Aedes aegypti* mosquitoes

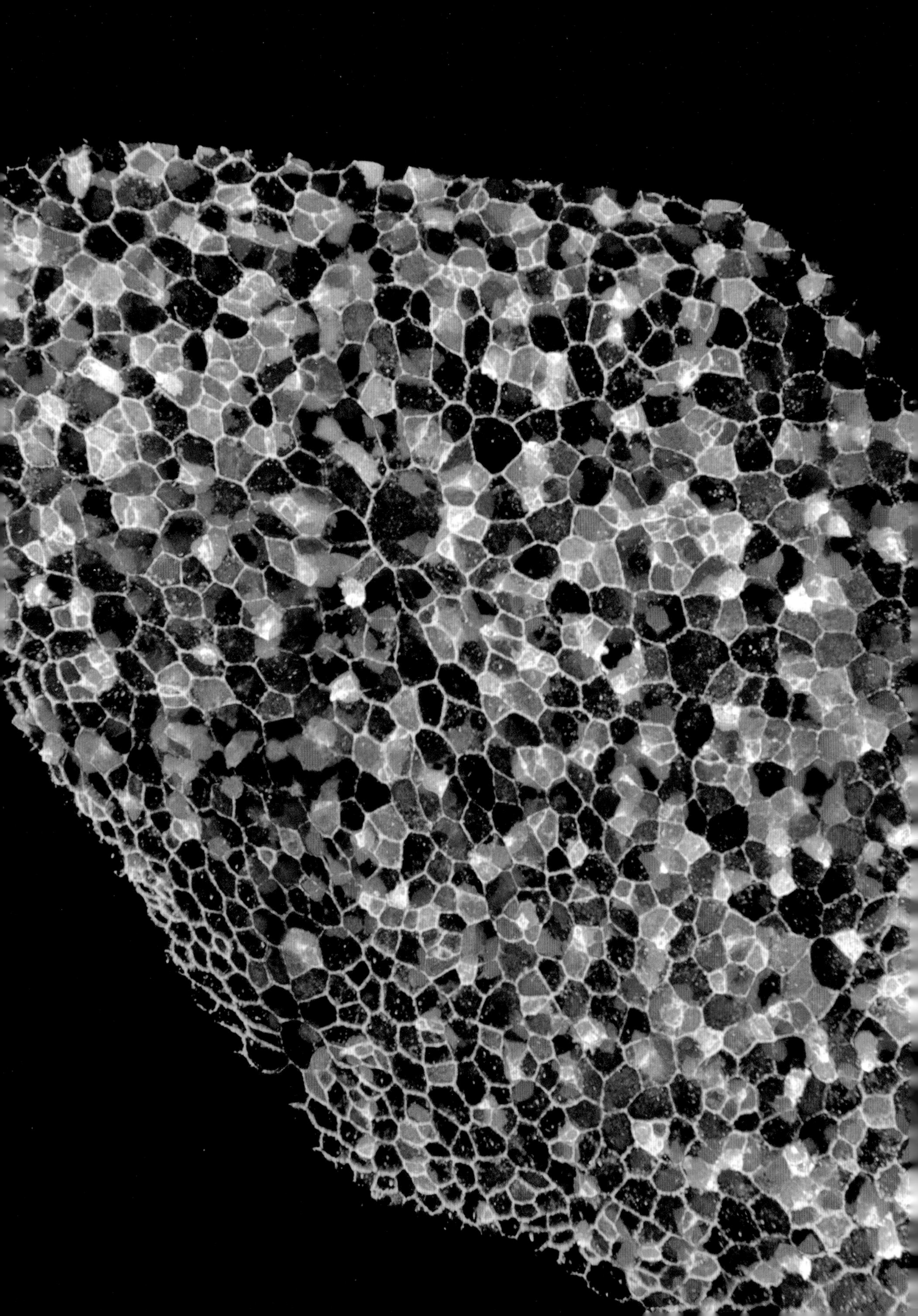

From Molecular Biology to Stem Cells

STEM CELLS
IN FIGURES

100,000 BILLION
cells in the human body, all derived from a single cell

200 TYPES
of cells produced from stem cells from various tissues and organs

10 YEARS
time required to replace the human skeleton; 4 months for a red blood cell; 2 weeks for a surface skin cell; 7 days for a corneal cell

May 1968... While students took to the streets of Paris, Jacques Monod and François Jacob were getting ready for what would prove to be a decisive turning point for the Institut Pasteur: the opening of a new department dedicated to molecular biology—in other words the study of the molecular mechanisms governing cell behavior. The venture, which included plans for purpose-built premises, had recently been approved by the Institut Pasteur Board of Directors and was fully in line with the discovery that had earned Jacques Monod, François Jacob and André Lwoff the Nobel Prize in 1965. The 1950s had seen the birth of a new understanding of how genes—sections of chromosomes that each carry a specific hereditary trait—govern the synthesis of proteins, the molecules that play a critical role in the function of living beings. The three Institut Pasteur biologists demonstrated the existence of mechanisms for regulating gene expression: they showed how proteins, in turn, regulate gene activity (see page 19). It was a natural progression for them to dedicate a department to the study of these regulation mechanisms, which had been discovered in bacteria but which, according to Jacques Monod, were valid in equal measure for bacteria and elephants.

Opposite: A layer of neural stem cells lining the brain ventricle of an adult zebrafish (*Danio rerio*). This image illustrates the heterogeneity and complexity of cell populations in the brain ventricle (orange: tight cell junctions, magenta: expression of the neural progenitor marker Ascl1, white: expression of the neural progenitor marker DeltaA).

François Jacob set to work on eukaryotes

When the department opened, most of the units were researching the regulation of gene expression. Several teams continued to explore this aspect in bacteria, but François Jacob was at last able to fulfill his long-harbored aspiration: to study these mechanisms in more complex organisms—eukaryotes, which are cells with a nucleus—and, in particular, animals, including humans. Jacob was convinced that mechanisms similar to those observed in bacteria were at work in eukaryotes and that they held the key to the mystery of embryonic development—the formation of a complex organism with several organs, each with its own functions and characteristics, derived from a single cell. How was this cell differentiation possible?

But this was an era with limited tools on offer for studying embryonic development and cell differentiation. It was possible to preserve an early mouse embryo in culture for several days, at a stage where it was only a few cells large, but this system was highly inadequate for studying what happens in tissue, for example. Nevertheless, a discovery made in the 1950s and 1960s captured Jacob's attention: two Americans—Leroy Stevens and Barry Pierce—selected a line of mice in which a particular type of tumor, known as a teratocarcinoma, developed spontaneously in the testicles. These tumors contain many different cell types—neurons, muscle cells, skin cells, etc., but without any organization. Stevens and Pierce showed that these tumors are transplantable: even a single tumor cell can give rise to a teratocarcinoma in another mouse. The tumor therefore contains cells that are able to differentiate into several different types just like embryonic cells. The two biologists had just proved the existence—in this case in a tumor—of what would later be called pluripotent stem cells.

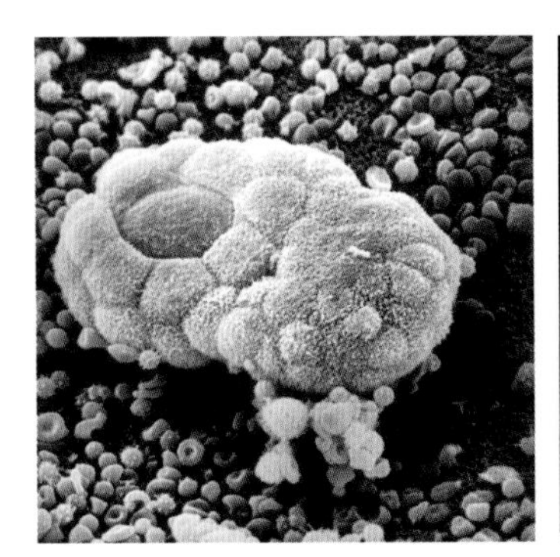

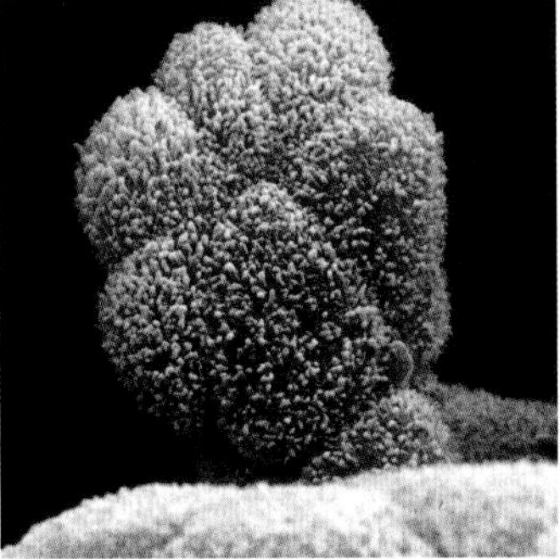

Above left: A teratocarcinoma, a malignant tumor used as a model for embryogenesis in mice.

Above right: A teratocarcinoma transplanted into a mouse peritoneum (development research).

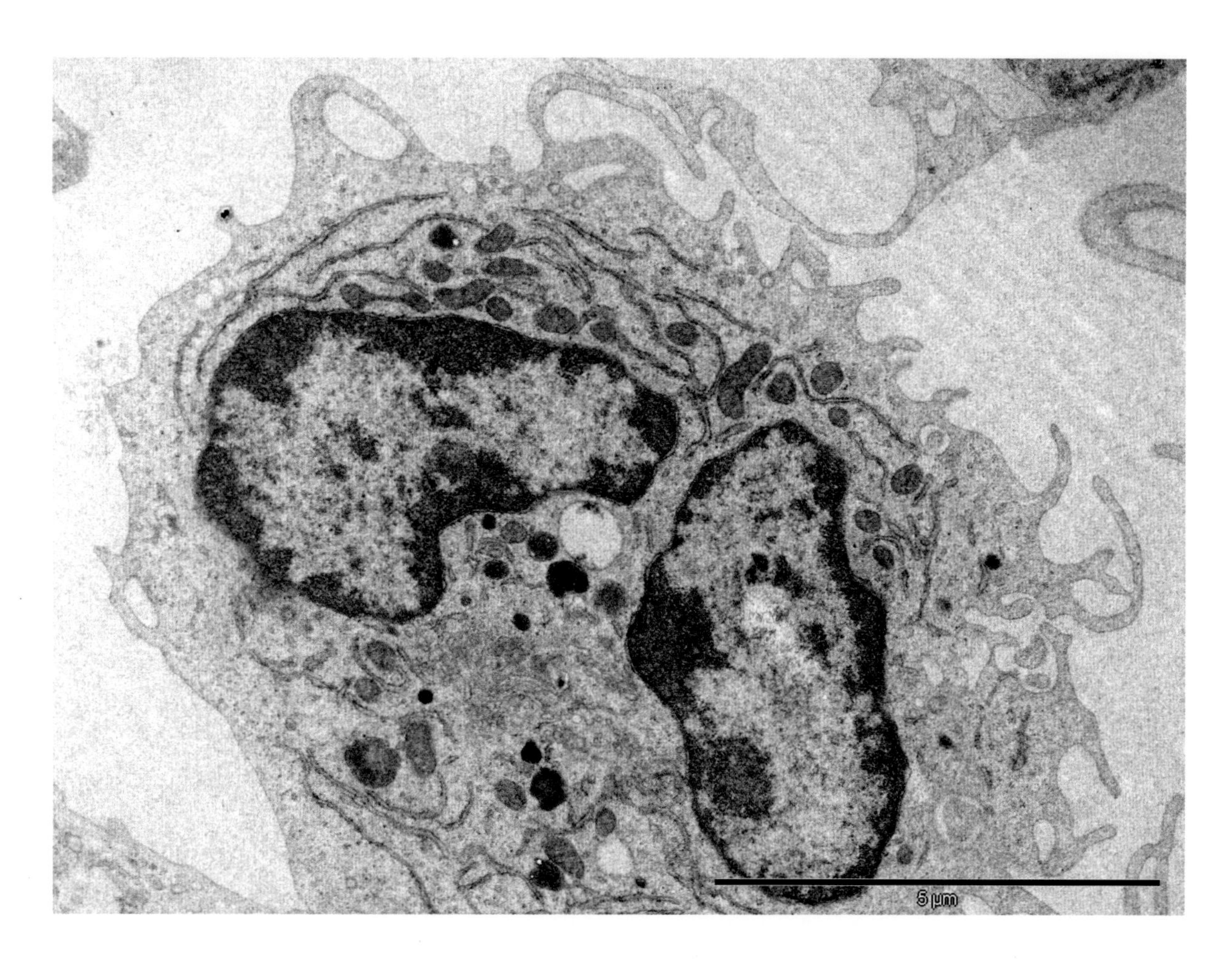

“The Department of Molecular Biology was a new departure for the Institut Pasteur because its line of research was completely unrelated to infectious diseases. François Jacob and Jacques Monod already had a vision of a different kind, with a focus on genes and how they were regulated in cells and tissues.”

Shahragim Tajbakhsh, Head of the Stem Cells and Development Unit

Section of a eukaryotic cell viewed using transmission electron microscopy. Scale bar: 5 microns.

Over time, embryologists such as Boris Ephrussi in the brand new Center for Molecular Genetics in Gif-sur-Yvette isolated these stem cells and showed their potential in the study of differentiation and the first stages of embryogenesis. Ephrussi passed on to François Jacob both his expertise and the cell lines he had produced. During the 1970s, Jacob's team—in particular Charles Babinet, Philippe Brûlet, Jean-François Nicolas, Philip Avner and Jean-Louis Guénet—, and then others in the UK and USA, established culturable cell lines that could produce all tissue types, including heart, muscle, nerve, and skin cells.

The early days proved difficult. Even though the team had a system for studying cell differentiation in culture, there were no tools to help them understand what takes place in these cells at the molecular level. François Gros, who joined the department with Margaret Buckingham, a young Scottish recruit, to focus on skeletal muscle-forming cells, faced the same problem. However, radical changes took place from the 1970s onwards, with the introduction of new molecular biology techniques. It became possible to isolate genes coding for specific proteins, to study their position on human and mouse chromosomes and their expression during differentiation, and even to begin studying the mechanisms for regulating their expression. Henri Buc's team carried out physicochemical and enzymology research and investigated protein-DNA interactions to elucidate the regulatory mechanisms of gene expression.

This research was complemented by the work of Daniel Louvard's team, which explored and promoted the discipline of cell biology within the department.

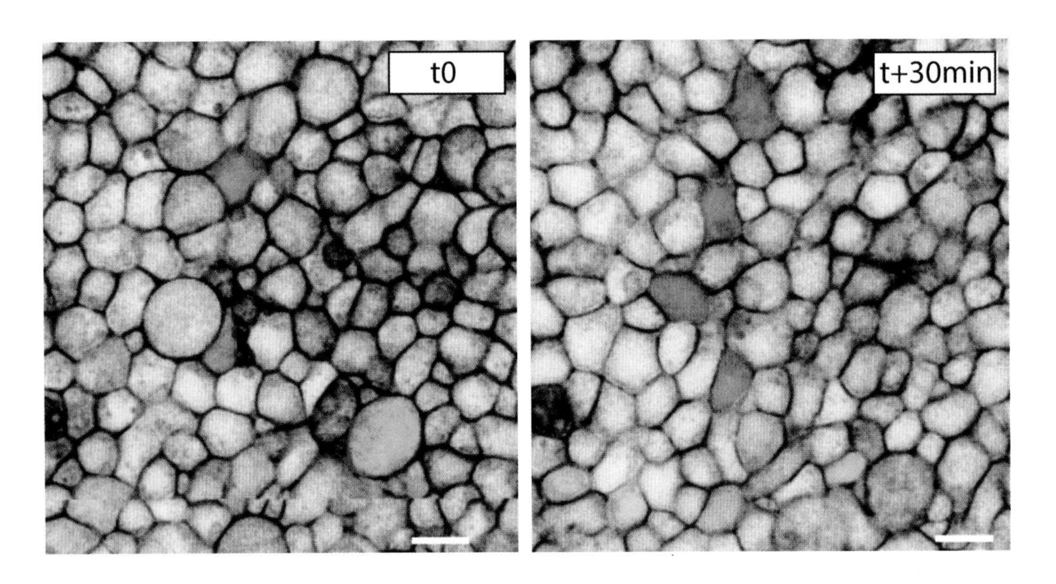

Daughter cells (cell pairs shown in the same color) from cell divisions in the early stages of embryogenesis, in chicken embryos.

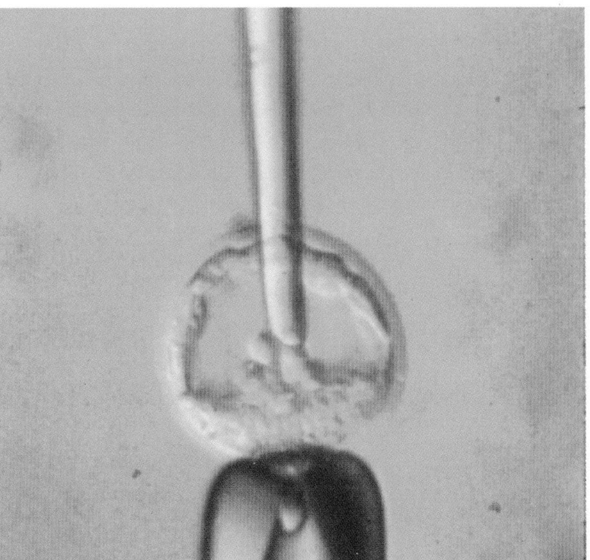

Meanwhile, research on gene regulation continued, with Anthony Pugsley and Bernard Dujon investigating single-cell models of bacteria and yeast, respectively. Their efforts supplemented work on gene regulation in more complex organisms. Yeast genetics also provided novel tools that subsequently became essential for genomics research in eukaryotes.

From the late 1980s onwards, several teams spearheaded fundamental developmental biology research. The teams led by Charles Babinet and Philippe Brûlet pioneered transgenesis and homologous recombination techniques in mice at the Institut Pasteur. Margaret Buckingham's team also used genetically modified mice to investigate the role of structural and regulatory genes in skeletal and cardiac muscle development, with a particular focus on embryology. The group led by Jean-Louis Guénet used traditional mouse genetics techniques to identify the loci and genes responsible for specific phenotypes. Jean-François Nicolas and his team initially studied differentiation in embryonic carcinoma cells, then used innovative molecular tools to mark and monitor specific embryonic cell lines. Philip Avner's laboratory led the way in identifying the mechanisms regulating X-chromosome inactivation and epigenetic regulation in mice. Together, these researchers explored fundamental questions in gene regulation and tissue organization. Their vital contributions helped consolidate a discipline that is still developing today.

Cells being injected into a mouse blastocyst (an embryo in the early stages of development).

Of flies, mice and men

At the Institut Pasteur the time had come at last to study development of the organism as a whole. Cell differentiation can only be understood if a study is carried out *in vivo*, in the organism as a whole, on the way cells interact and cooperate to form different tissues according to an established plan. This led to a surge in developmental research worldwide.

Firstly, since the start of the 1980s, thanks to the work of Martin Evans and his colleagues at the University of Cardiff (UK), it was now possible to extract stem cells from the embryos of healthy mice and culture them in their undifferentiated state. The team had even succeeded in genetically modifying stem cells to produce mice. These breakthroughs led to the development of transgenic mice—an effective tool for developmental biologists. Using this technique, Mario Capecchi from the University of Utah produced mice with a target gene deleted, or 'knocked out'. Knockout mice soon became widespread in laboratories (homologous recombination technology) and earned their co-inventor Oliver Smithies the Nobel Prize in 2007, alongside Martin Evans.

Secondly, scientists had started to describe the genes involved in development. Edward Lewis at the California Institute of Technology, Christiane Nüsslein-Volhard and Eric Wieschaus from the European Molecular Biology Laboratory (EMBL) in Heidelberg (Germany) and their colleagues identified a number of genes in flies that play a vital role in development. This research, carried out at the end of the 1970s, received the Nobel Prize in 1995. In 1983, the teams of Walter Gehring, from the University of Basel (Switzerland), and David Hogness, from the University of Stanford (USA), isolated and characterized these genes, whose mutations were responsible for malformations that

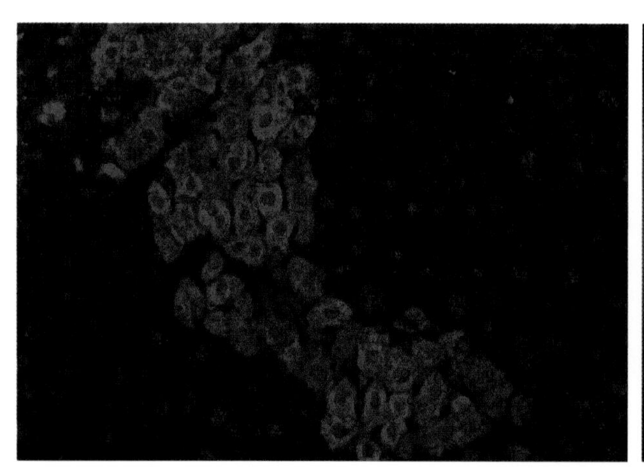

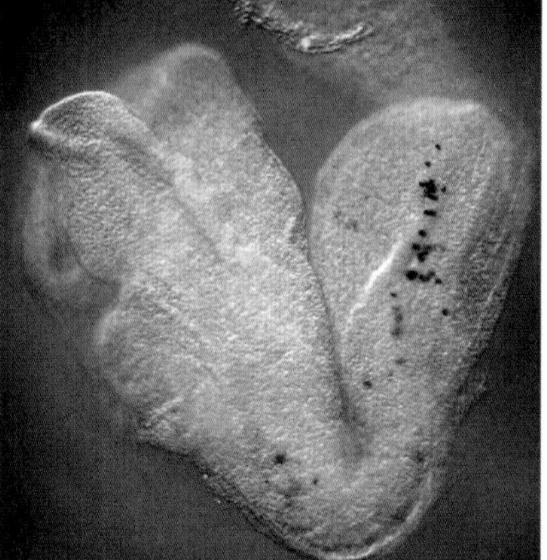

Above left: Human iPS cell differentiation to hepatocytes (in red) in a section of a mouse liver. The cell nuclei are visible in blue.

Above right: In blue, cells from a mouse embryo deriving from a bipotent stem cell.

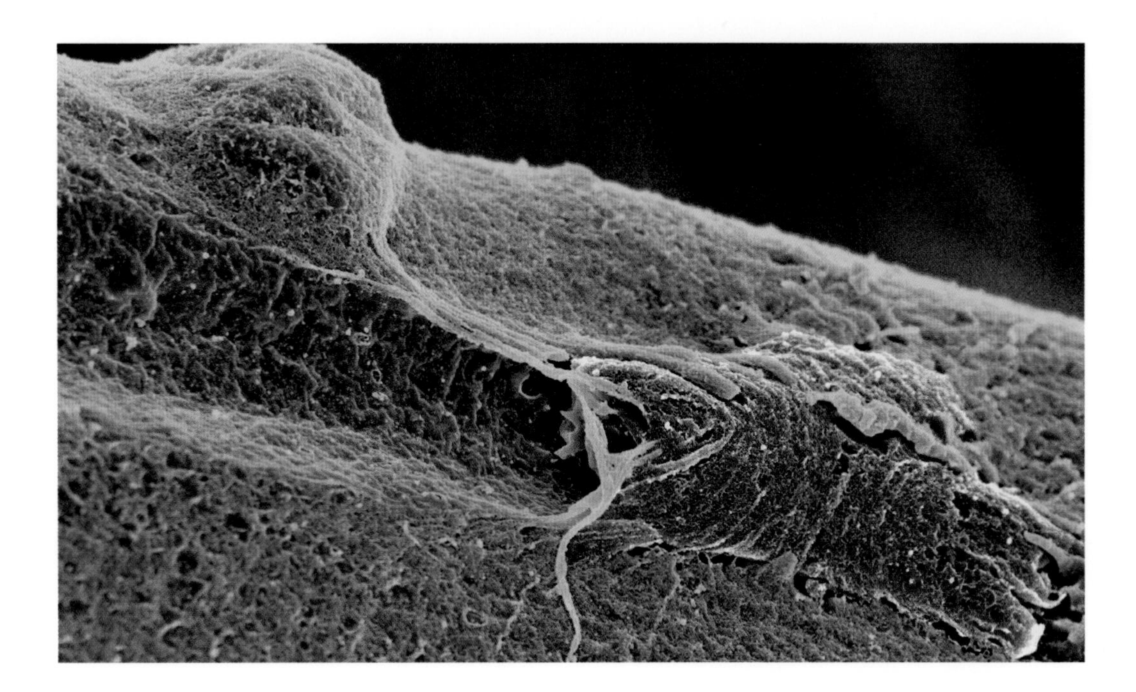

had been described in flies since the beginning of the 20th century. Homologs of these genes were subsequently found in other animal lines such as mice and humans, proving that the mechanisms involved had been well conserved in the animal kingdom.

In addition, a fundamental discovery was made: various teams, notably the team led by Takashi Matsui of Washington University in St. Louis (USA), showed that, in eukaryotes too, certain proteins play a part in the regulation of gene expression. Known as transcription factors, these are particularly instrumental in cell differentiation, guiding the DNA transcription machinery towards the genes to be transcribed into proteins. For each cell type to be derived from stem cells there is a defined cocktail of transcription factors that activate the expression of genes specifically for it. In the 1990s-2000s, hundreds

SKELETAL MUSCLE STEM CELLS

Shahragim Tajbakhsh and his colleagues identified a population of muscle stem cells that were in a more quiescent, or dormant, state than the others but capable of becoming active in the event of injury—a sign that they can change their metabolism. In collaboration with Fabrice Chrétien's team, the groups showed in both mice and humans that these cells can survive for more than two weeks *post mortem* and retain their regenerative potential.

A skeletal muscle stem cell (in yellow), in activation state, on a muscle fiber (in blue).

of transcription factors were identified as associated with a particular tissue type—bone, skin, cartilage, muscle, etc. The Institut Pasteur played a part in this characterization process. Particularly worthy of note in the 1990s was the work of Moshe Yaniv, Mary Weiss and their colleagues, who showed that these proteins regulate liver cell differentiation in mice.

Studying stem cells in their own right

We now know both how to differentiate a stem cell into any cell type and to restore a differentiated cell to a pluripotent state. The technique for producing "induced" pluripotent stem cells (iPS cells) is much simpler to implement than transferring a cell nucleus into an oocyte (the technique used to clone Dolly the sheep in 1997), and was developed by Shinya Yamanaka of Kyoto University (Japan) in 2006, and rewarded by the Nobel Prize in 2012. Biologists also have access to tens of thousands of modified mice, some carrying several genes mutated in several places or with a deleted gene. It is also possible to produce fluorescent mice: a gene of interest is simply tagged with a jellyfish gene coding for a fluorescent protein. When this sequence is expressed it produces a protein carrying a fluorescent marker that shows up under the microscope—during embryogenesis, for example–and in the living organism.

> "Progress in pluripotent stem cell research has enabled us to generate mini-organs in culture."
>
> Shahragim Tajbakhsh, Head of the Stem Cells and Development Unit

Opposite top: Human induced pluripotent stem cells examined under a microscope.

Opposite, bottom left: Bone marrow stem cells specializing into two distinct cell types, visible on the photo: dendritic cells (in green) and macrophages (in red). The cell nuclei are shown in blue.

Opposite, bottom right: A layer of neural stem cells lining the brain ventricle of an adult zebrafish (*Danio rerio*).

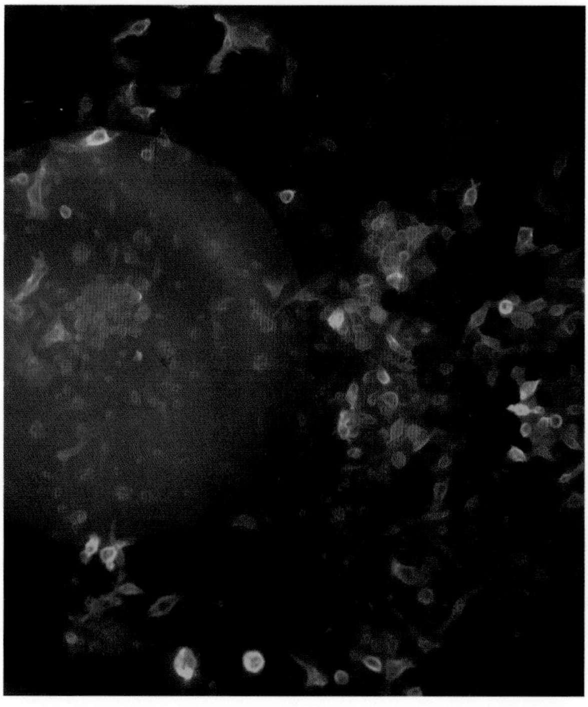

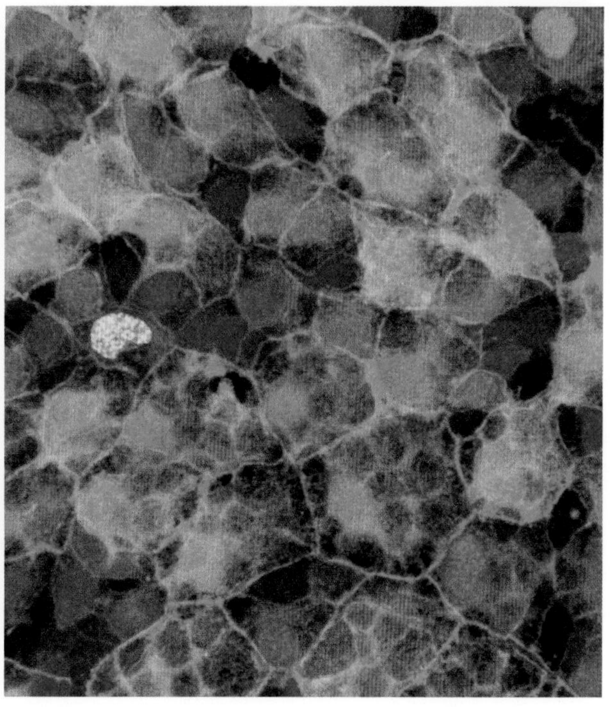

THE PROMISE OF STEM CELL RESEARCH

This work has a number of important goals: not only to gain an understanding of the fate of embryonic stem cells and the formation of tissues and organs, but also to explore the biology of stem cells in the hope of using them in regenerative medicine for diseases such as Parkinson's, Alzheimer's, diabetes, certain muscle disorders or retinal diseases such as ARMD (age-related macular degeneration)—which are all associated with progressive cell death. Clinical trials are under way to treat ARMD patients: a few skin cells are taken and reprogrammed as induced pluripotent stem cells; the cells are then differentiated into retinal precursor cells and grafted onto the patient's eye. The initial results are promising, but it will be a while before the technique can be used routinely.

A scientist in the Stem Cells and Development Unit in March 2017.

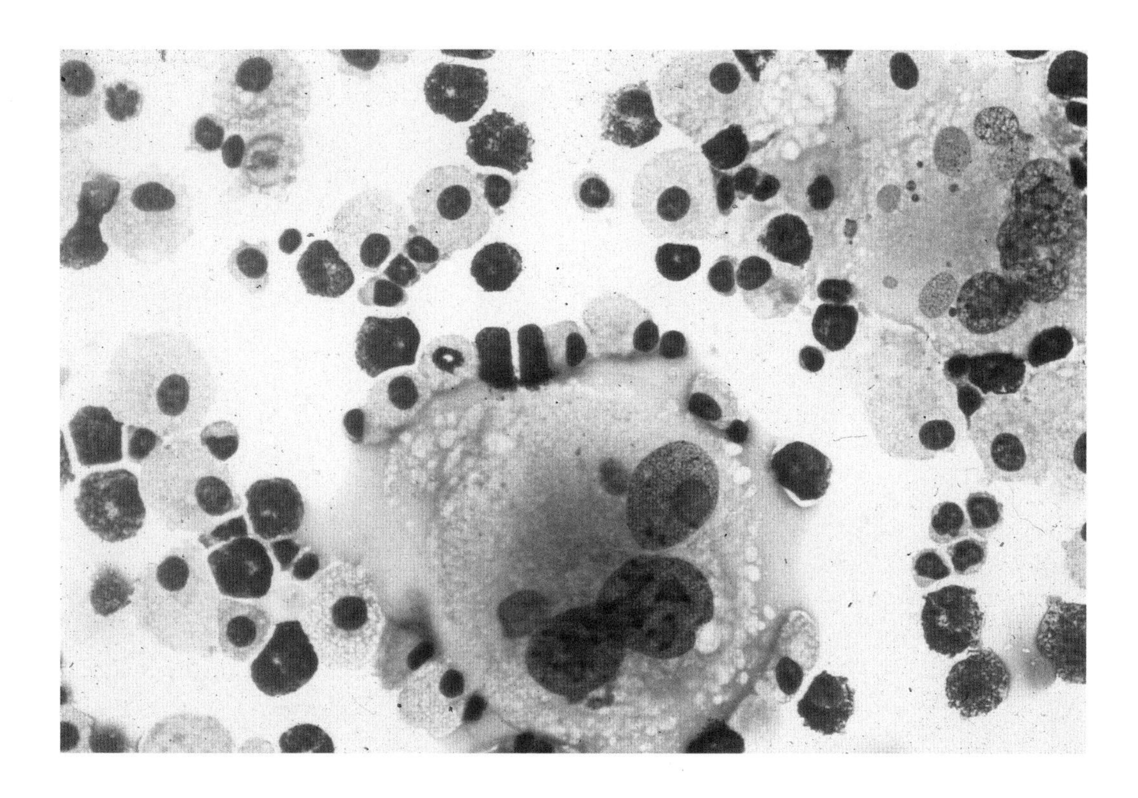

Biologists work with all these tools, not only to identify the mechanisms at play during embryo development, but also to achieve more accurate characterization of stem cells themselves.

Generally, it is vital to make sure that pluripotent cells are eliminated, to avoid the formation of tumors. It is also important to determine whether these artificially produced pluripotent stem cells have the same functional potential as embryonic stem cells. As long as this question remains unanswered it is unlikely that they will be used in cell therapy. However, this would solve an ethical dilemma, given that embryonic stem cells are taken from surplus embryos that were frozen following *in vitro* fertilization for infertility. Induced pluripotent stem cells would be produced from the patient's own tissues, and so also help to reduce transplant rejection.

A fusion of several stem cells—or a myotube—obtained *in vitro* from a human muscle collected 17 days after an individual's death. Cell properties are also maintained *in vivo*.

The abilities of stem cells

A major advantage of using a variety of model organisms is that scientists can focus on the salient features of each. François Schweisguth's team, for example, has used live-cell imaging of sensory organ precursor cells in flies to examine cell fate determination in space and time. This is not a typical stem cell system involving self-renewal; instead, several rounds of asymmetric cell division yield multiple cell types including a neuron and associated cells. A number of regulatory factors including Notch and Numb are involved in these asymmetric cell divisions. This fundamental research provides critical insights into asymmetric cell division, a mechanism used by stem cells in many organisms to generate a variety of cell fates and to self-renew.

Pablo Navarro Gil and his team are attempting to identify the mechanism that enables pluripotent stem cells to be passed from one generation to the next. Embryonic stem cells renew themselves—in other words some of them generate new stem cells to form a cell reservoir.

It is thanks to these cell reservoirs that some animals, such as lizards and salamanders, are able to regenerate limbs even in adulthood. Mammals, however, are not so lucky, and must repair any injuries they incur. Adult mammals retain some stem cells, but most of these lose their regenerative potential with age. For example, hematopoietic stem cells residing in the bone marrow continually renew red blood cells, which have a short life span. Used for over fifty years to help treat certain blood and immune system disorders, the stem cells are transplanted by IV injection and migrate to the host's bone marrow, where they differentiate correctly. Biologists are trying to unlock the secrets of these stem cells: What is their role? How does aging affect them? Could they be reactivated?

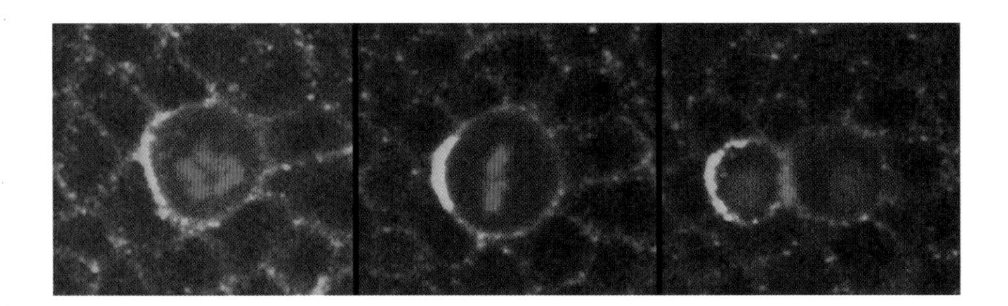

Above: This image shows the asymmetric division of a sensory organ precursor cell (chromosomes and plasma membrane in red) in a developing fly. A protein that regulates cell fate (Numb, in green) segregates to one pole and is inherited by just one of the two daughter cells.

Opposite: Multipotent stem cells can trigger selective differentiation using specific inducers.

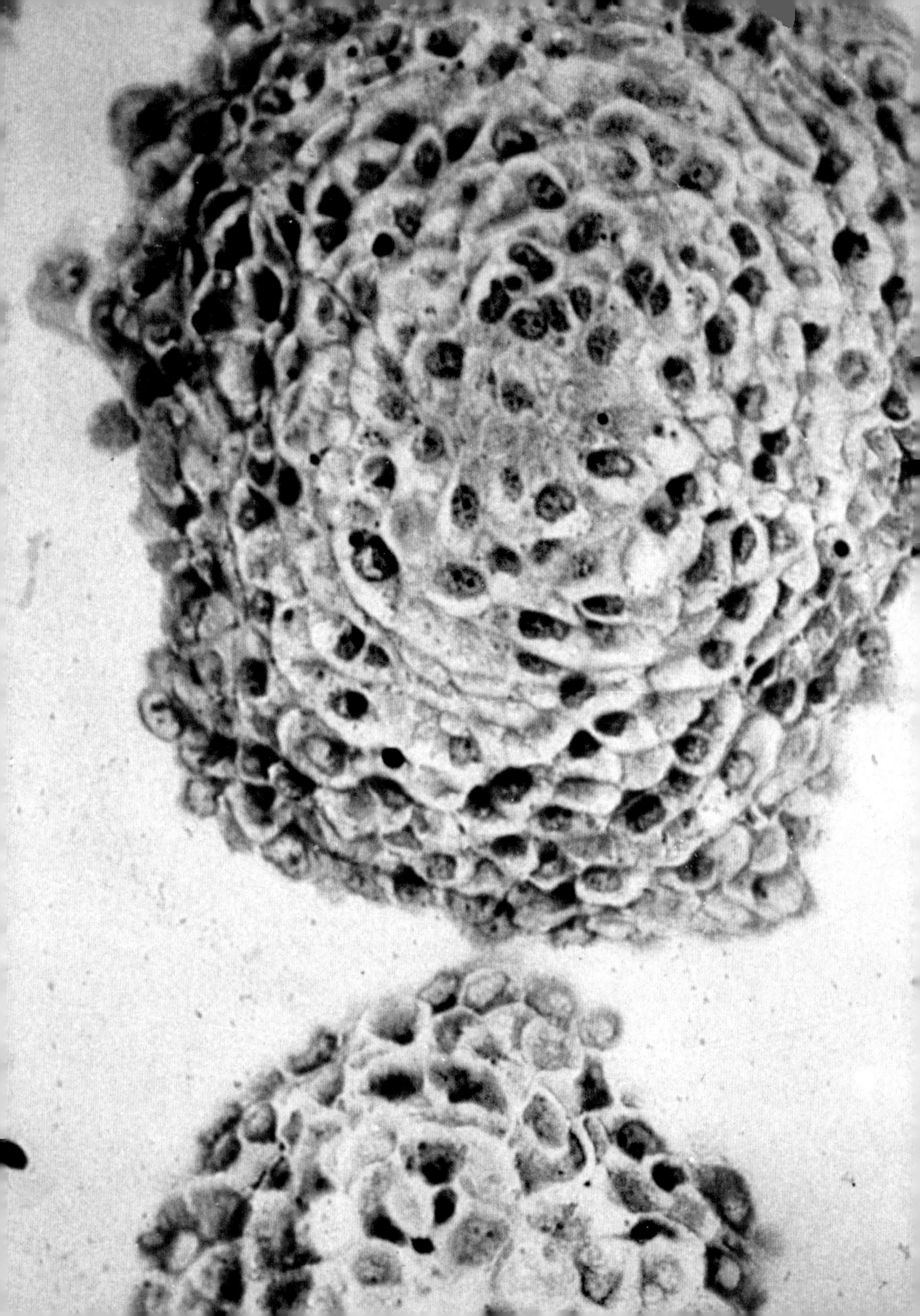

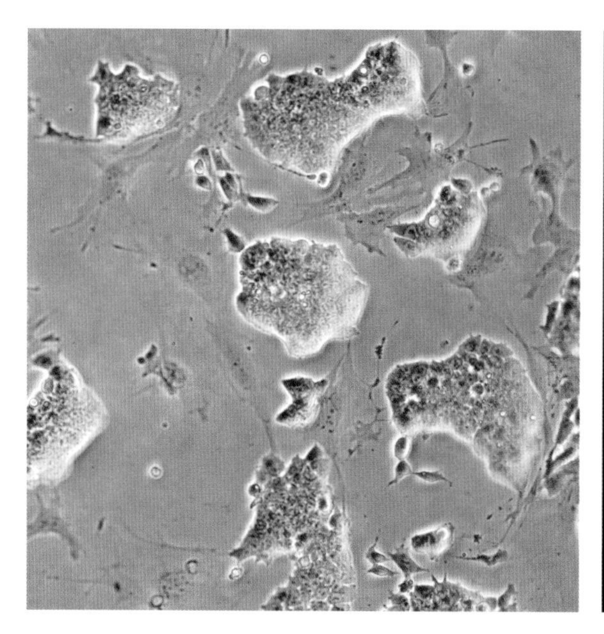

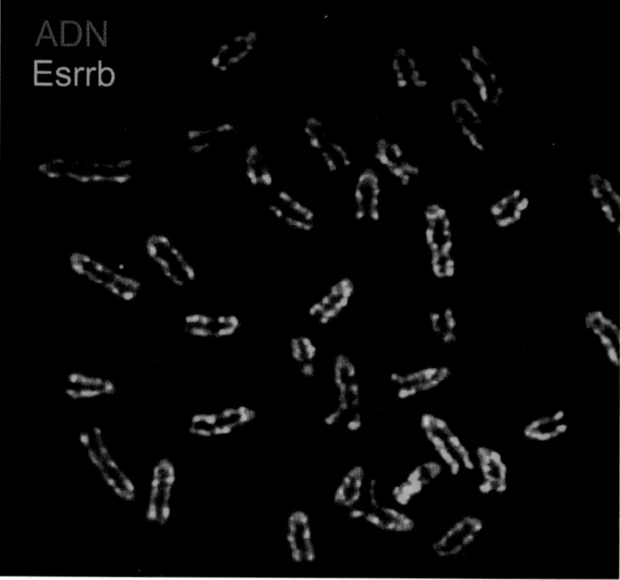

During the last few years, the team led by Philippe Herbomel within this Institut Pasteur department used zebrafish to identify the niche where hematopoietic stem cells are produced. Using a molecular tracer he was able to monitor *in vivo* how these cells emerge in the animal model.

Although stem cell research is a major focus of interest at the Institut Pasteur, it is not an end in itself, as illustrated by this brief overview of the changing nature of research topics in what is now known as the Department of Developmental and Stem Cell Biology. Understanding the fundamental principles of gene regulation and cell fate involves a wide variety of different disciplines, including epigenetics, epigenomics, biophysics, single-cell analysis and real-time imaging, all of which converge to shed light on how interactions between cells result in organ morphogenesis during growth, regeneration, illness and aging. The major strength of this department is its sheer diversity—from the wide range of model organisms (flies, worms, zebrafish, mice and birds) and sophisticated techniques it uses to its multiscale research at molecular, cell and tissue level. The recent recruitment of a new generation of talented leaders with expertise in all these fields will further consolidate this diversity.

These are only a few of the research areas being investigated in this field at the Institut Pasteur. François Jacob and Jacques Monod could not have imagined a brighter future for their molecular biology department!

Above left: Mouse embryonic stem cell karyotype stained for the Esrrb transcription factor.

Above right: Embryonic stem cell colonies with rare cells isolated during spontaneous differentiation.

A TITAN AT THE INSTITUT PASTEUR

At the Institut Pasteur it will soon be possible to observe biological molecules in their natural environment and with near-atomic resolution. The final addition to the Center for the Analysis of Complex Systems in Complex Environments (CACSICE), a Titan Krios™ electron microscope, is due for installation in late 2017. CACSICE is coordinated by the Institut Pasteur, in cooperation with Paris-Descartes and Pierre et Marie Curie Universities and the Physical and Chemical Biology Institute (IBPC) in Paris. The center already has a vast range of equipment for probing the structure of biological molecules, such as X-ray crystallography, liquid- and solid-state nuclear magnetic resonance (NMR), small-angle X-ray scattering and structural mass spectrometry. All these techniques are essential for studying various aspects of molecular structure and function, but cannot be used to observe molecules in their natural environment. The Titan Krios™ is a cryo-electron microscope that reveals the molecule in a state that is closer to its natural state than can be achieved by producing a crystal of the molecule in question, as in crystallography. Michael Nilges, who heads up the Structural Biology and Chemistry Department and coordinates the CACSICE project, already foresees multiple applications: the study not only of proteins, but also of macromolecular complexes such as ribosomes—cellular protein builders—and even subcellular compartments or whole viruses.

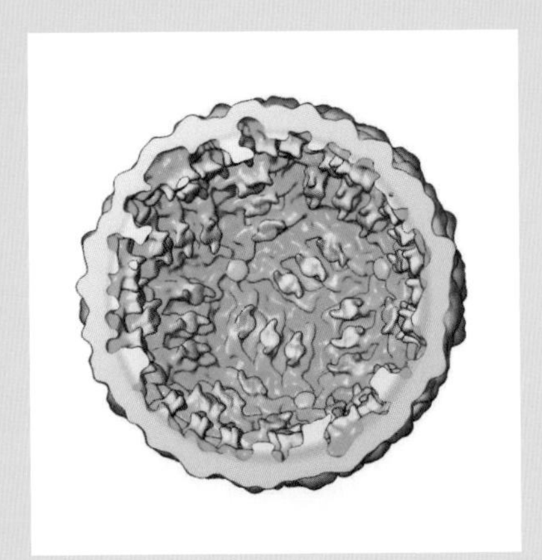

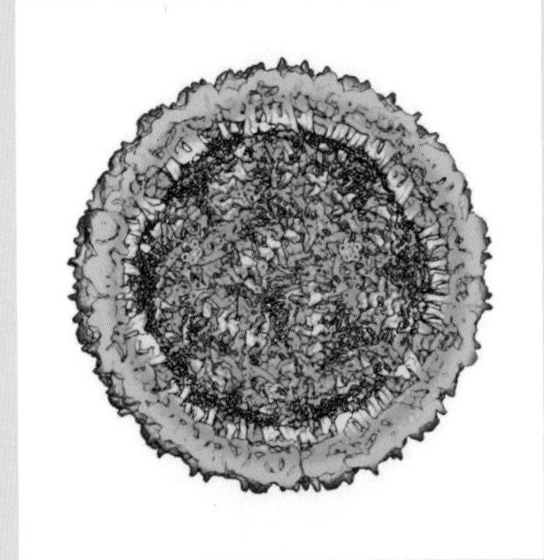

Above left: The dengue virus in standard electron microscopy.

Above right: The dengue virus observed under a Titan microscope.

Gut Microbiota

Key to Human Health

THE GUT MICROBIOTA IN FIGURES

100,000 BILLION
(or 10^{14}) microorganisms live in our digestive system—that's ten times more than the number of cells in our entire body!

A BACTERIUM IS 1,000 TIMES
smaller than a human cell

3.3 MILLION
genes identified as belonging to more than 1,000 microorganism strains make up the gut microbiota. By comparison, the human genome contains around 35,000 genes

1.5 TO 2KG
is what the adult gut microbiota weighs

By the late 19th century, Louis Pasteur's discoveries had branded microbes as public enemy number one. They were responsible for severe infections and had to be brought under control, whatever the cost. A hundred and fifty years later, the media and best sellers are telling us that the microbes living in our intestines are "friendly" and good for our health! So maybe it's time we took a closer look...

Opposite: *Bacteroides fragilis*. Strict anaerobic bacteria found in natural cavities in humans (e.g. the digestive tract), responsible for post-operative secondary septicemia or gynecological infections.

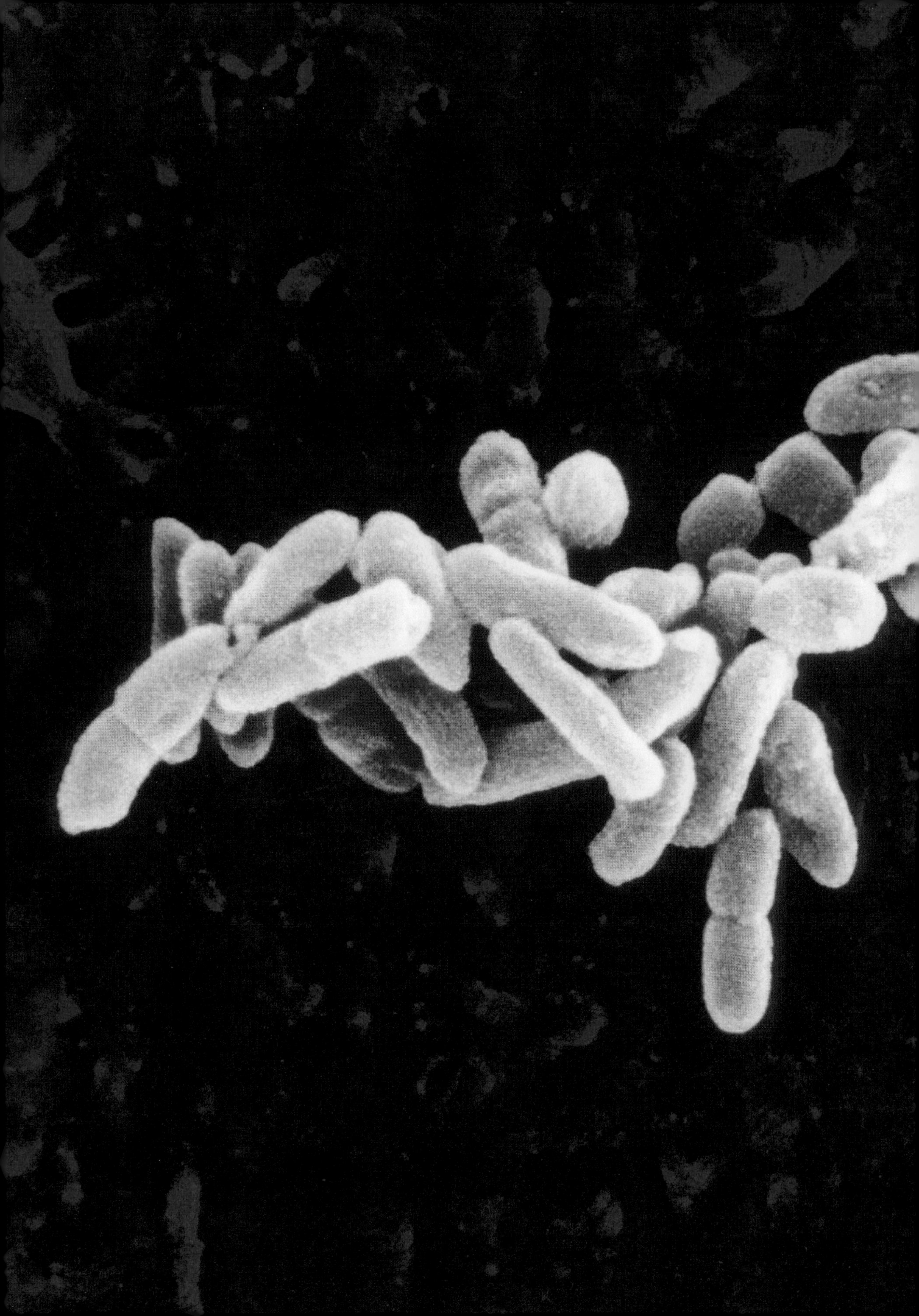

Previously known as gut flora, the gut microbiota is made up of thousands of billions of bacteria, yeasts and viruses that naturally occur in our intestines. This "organ" is neither palpable nor viewable using medical imaging techniques, yet it weighs more than a kilogram and we couldn't survive without it! Many of our body's functions depend on symbiosis with the microbiota—and deregulation of the microbiota is linked to a host of disorders ranging from obesity to Parkinson's disease and including some forms of autism.

Symbiosis is an association between two or more different species of organisms.

The Russian scientist Ilya Mechnikov, who joined the Institut Pasteur in 1888, was the first to advance the theory that the microbes living in our intestines might have a positive influence on our health. His hunch went against the prevailing view of the time that microbes were our enemies. After observing the longevity of Bulgarian populations—avid consumers of kefir (fermented milk), many were living to be more than a hundred—Mechnikov even suspected that some microbes in food might interact in a beneficial way with our own microorganisms. He also contended that the microbiota was responsible for aging. Mechnikov's wife Olga, also a scientist at the Institut Pasteur, focused her research on the influence of bacteria on animal development. Between the two of them—especially when Mechnikov became Deputy Director of the Institut Pasteur in 1904 and was awarded the Nobel Prize in 1908 for his research on immunity (see page 14)—they inspired many scientists to come and study the human microbiota in Paris. And today it is still one of the research topics under investigation by teams at the Institut Pasteur, especially those directed by Philippe Sansonetti and Gérard Eberl.

Louis Pasteur in the microbe culture chamber at the École Normale Supérieure (drawing by Poyet, published in the journal *La Nature* in 1884).

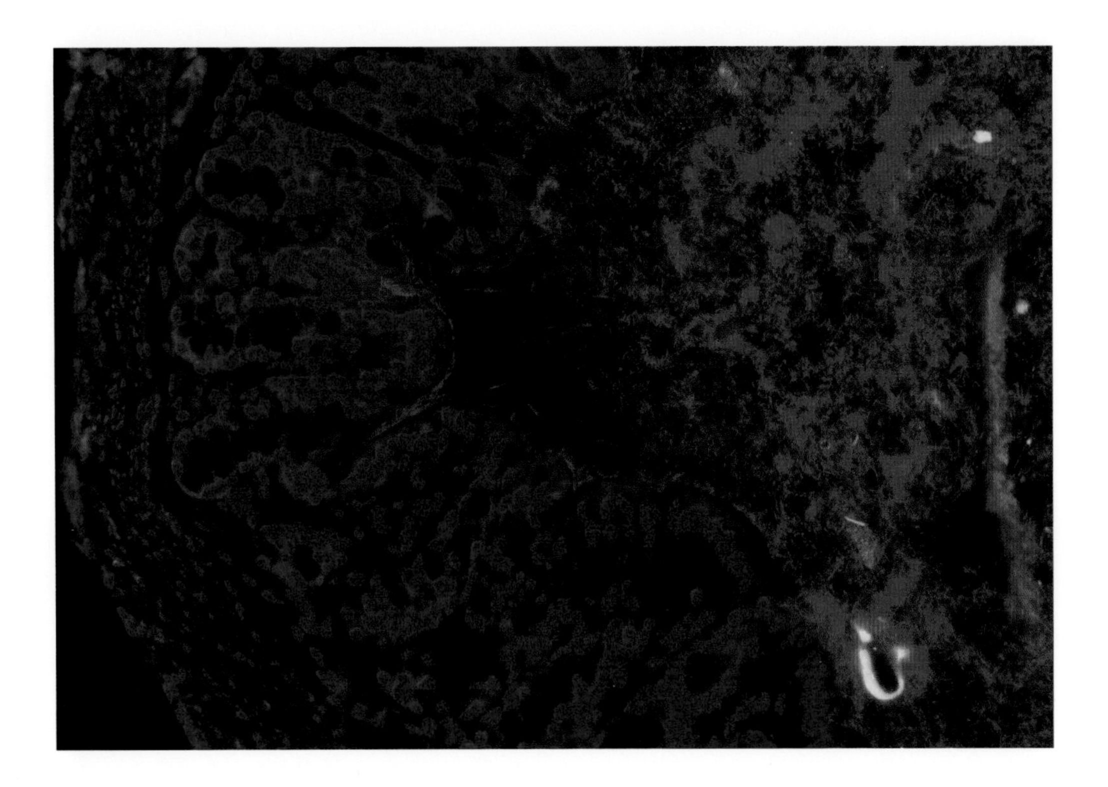

Molecular biology is the field that investigates how organisms work by examining their molecules, especially their genes.

Prebiotics are food components or molecules that contribute to the growth and/or activity of some bacteria in the microbiota.

Probiotics are living microorganisms that can be found in fermented foods (cheese, sauerkraut, bread, beer, etc.) and have a beneficial effect on our health.

From fetus to adulthood

Gut microbes are very hard to culture in laboratories—making them difficult to study—because they naturally live in the absence of oxygen (they are anaerobic) and often feed on little-known foods. We have only been able to improve our knowledge because of progress in molecular biology and metagenomics. Recent studies have shown that in healthy adults, the composition of the microbiota is relatively stable. It contains hundreds of different species of microorganisms, although two groups of bacteria, *Firmicutes* and *Bacteroidetes*, are by far the most abundant. This biodiversity, together with the actual workings of the microbiota, is influenced by external factors such as food (mainly its fiber, prebiotic and probiotic content) and also antibiotic treatments that can cause temporary disruption, in some cases having an impact on our health.

A mouse colon with the cell nucleus marked in blue, the cells themselves in green and the bacteria in red.

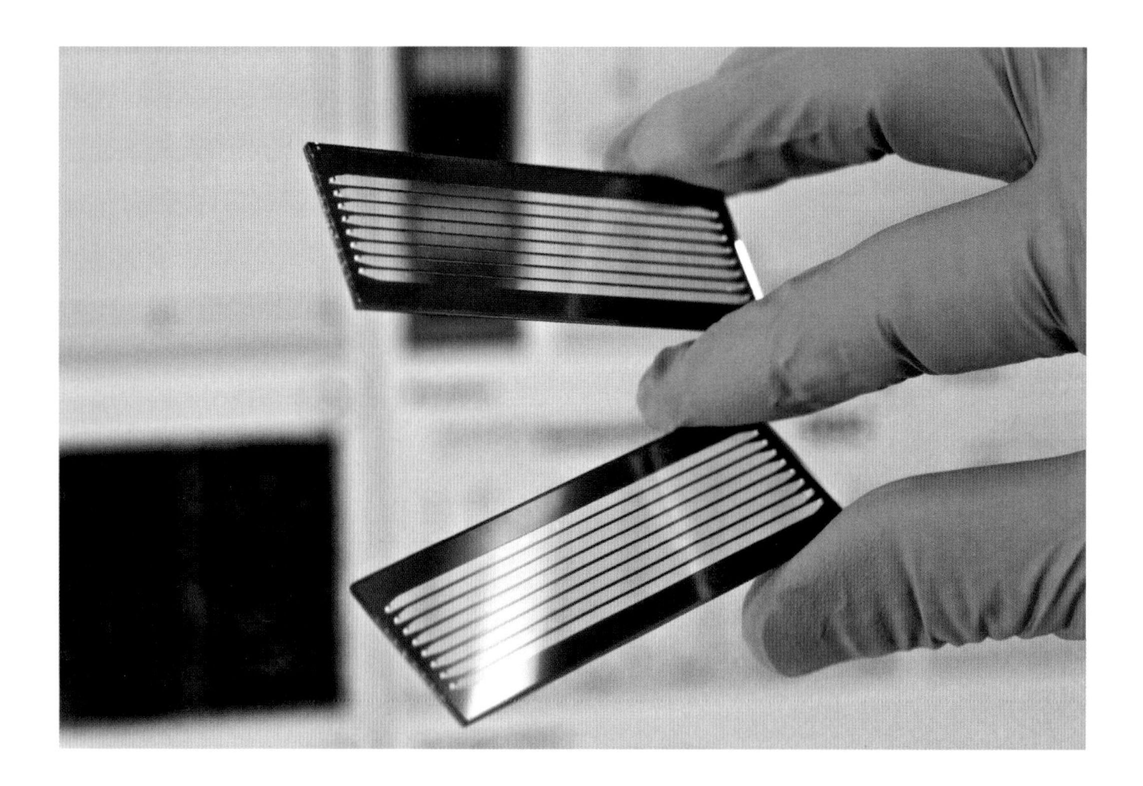

METAGENOMICS, A TECHNOLOGICAL REVOLUTION

Metagenomics, which combines high-throughput sequencing and big data analysis, has been applied to a host of different fields in recent years, including oceans, soil, and even our intestines. This technique can be used to characterize microbial diversity and provide an exhaustive survey of a large number of microorganisms that it would be difficult—if not impossible—to isolate and culture in laboratories.
The first stage involves taking a sample, for example from human feces or soil found in a meadow, then isolating all the DNA in the sample using physicochemical or enzymatic methods. This results in a "bulk" set of DNA from all the bacteria, yeasts and viruses in that environment.
The second stage is high-throughput sequencing: nowadays we have machines that are capable of sequencing tens of billions of nucleobases each day. Then, using big data analysis software, these data are processed: the DNA sequences are compared with each other and with other known sequences to identify genes and advance hypotheses about the workings of the microbial sample under study.
Sophisticated metagenomics techniques have made it possible to precisely characterize the biodiversity of the human gut microbiota and to understand how it interacts with the rest of the body. These techniques have been used to analyze seawater samples from the entire planet and to identify more than 40 million genes in marine plankton, 80% of which were previously unknown. And they have also resulted in the discovery of potential new antibiotics in soil samples and enzymes capable of breaking down environmental pollutants.

Sequencing: sequencing a DNA fragment means determining the order of the nucleobases in the fragment, resulting in a long string of letters. The human genome is composed of 3 billion nucleobases.

Slides for high-throughput sequencing. Two flow cells for the HiSeq Illumina sequencer.

Environmental factors also have a considerable influence on the establishment of the gut microbiota during the first two to three years of life. While babies are still in the womb, they are considered to be sterile. It is only when they are born, if the mother has a natural birth, that they come into close contact with the mother's vaginal and intestinal microbiota. This is a crucial stage in the constitution of the baby's own microbiota, which can be disrupted in the event of a cesarean delivery. Even in adulthood, differences in microbial biodiversity can still be observed between individuals born by natural childbirth and those born by cesarean section. If the baby goes on to be breastfed, he or she will regularly ingest the mother's skin microbiota and receive probiotic bacteria and prebiotic sugars from the breast milk.

From digestion to protection against pathogens

One of the main functions of the gut microbiota is to break down complex foods that our body is unable to digest by itself: plant fibers such as pectin found in fruit and vegetables, and resistant starch in cereals. Several bacterial species are involved in this process in turn, transforming these foods into nutrients that they can use themselves and/or into substances that can be absorbed by our body. In so doing, they also produce molecules that are vital for us, such as short-chain fatty acids (acetic acid) and vitamins B and K. If we do not ingest enough nutrients, or if they are not varied enough, the entire microbiota gets out of balance: its biodiversity is reduced and its function is impaired. This can lead to disruptions in the symbiosis between the microbiota and the rest of the body.

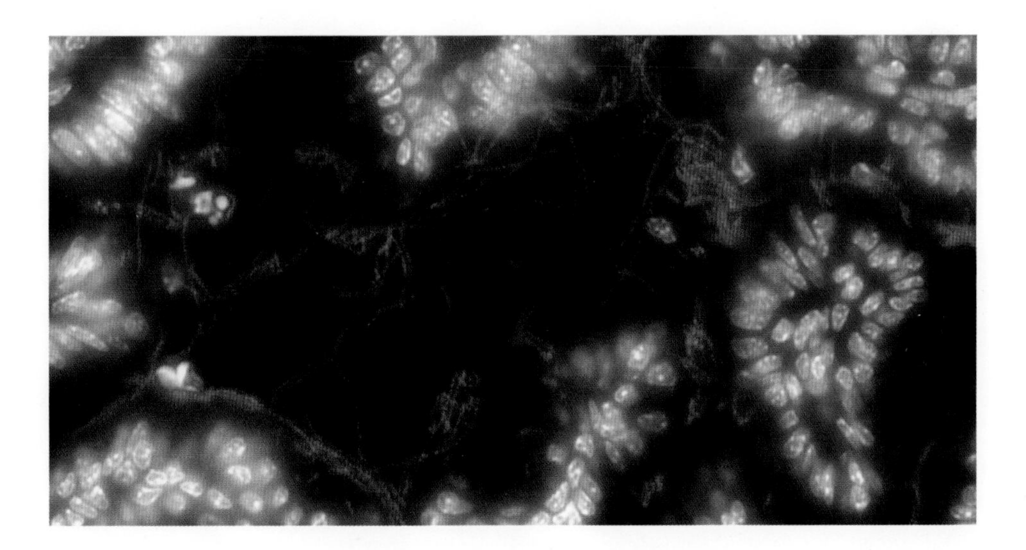

Symbiotic bacteria (in red) attached to the intestine (in white).

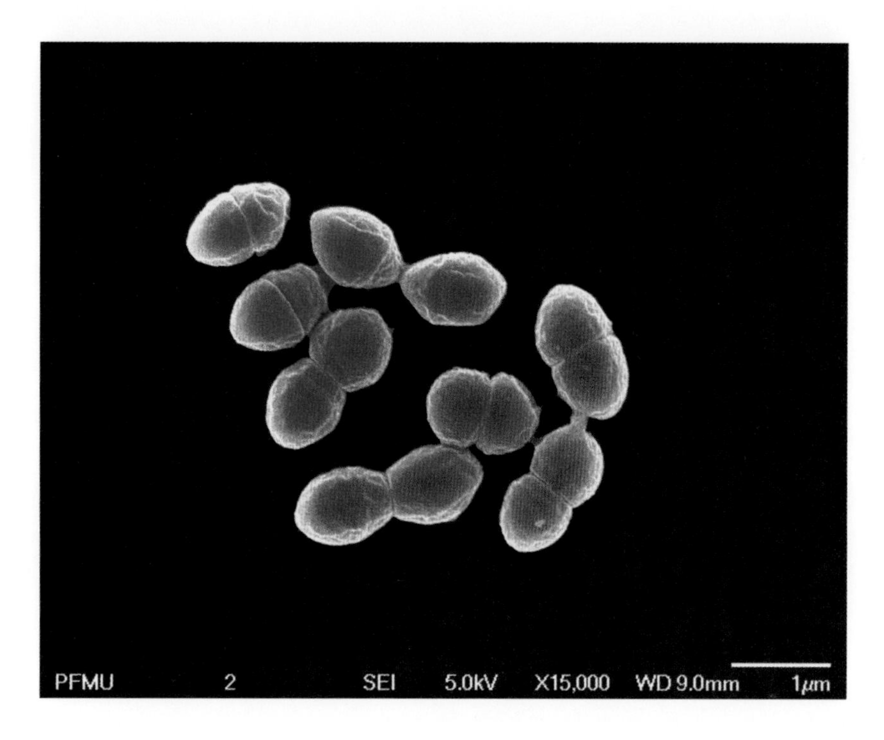

Another major role played by our intestinal microbes is protecting us from pathogens—firstly by simply taking up space, but also, more importantly, by maintaining an ongoing dialog with our immune system. When we are born, our immature immune system is unable to tell the difference between dangerous microorganisms and those that are good for our health. As the gut microbiota gradually develops, it educates our immune system. Research using mice models has demonstrated that if they are lacking a microbiota, they have fewer T lymphocytes and antibodies, both of which play an important role in immunity, and they also have some anomalies in immune organs such as the spleen and the lymph nodes. Moreover, by naturally producing molecules such as lipopolysaccharides, the microbiota maintains a degree of local inflammation, a phenomenon that is vital to keep our immune system permanently on guard. We also know that the microbiota has an influence on the way our body reacts to vaccination or to cancer treatments such as immunotherapy.

5

Immunotherapy is an approach that uses the properties of our own immune system to treat a disease by stimulating specific components or inhibiting inappropriate reactions.

But that is not all the microbiota does. There are currently increasing numbers of studies on this topic, and some have already revealed that the symbiosis between our body and our intestinal microbes influences a wide variety of functions including hunger regulation, growth, and sensitivity to pain and stress.

Opposite: Gut microvilli in mice viewed using scanning electron microscopy.

Top: *Enterococcus hirae*, a bacterial species in the human body, viewed using scanning electron microscopy.

Maintaining a healthy balance

The discovery and gradual understanding of the intricate relationships involved in this symbiosis, which plays such a vital role in the proper functioning of our body, are leading today's scientists, such as Gérard Eberl, Head of the Microenvironment and Immunity Unit at the Institut Pasteur, to view the entity formed by our organism and its microbes as a "superorganism", in other words an extremely well organized "microsociety".

Disruption to the composition and/or workings of the gut microbiota is referred to as dysbiosis. In recent years, several studies have demonstrated dysbiosis in people with chronic inflammatory bowel diseases (IBDs) such as Crohn's disease: compared to healthy individuals, these people have more bacteria causing inflammation and fewer anti-inflammatory microbes.

Links between dysbiosis and obesity and/or diabetes have also been demonstrated. We know that the microbiota in people with these conditions is different from that of healthy individuals. If we transplant an "obese" microbiota into a mouse of average weight with no existing gut microbes, it will go on to develop obesity and diabetes. But it has also been shown that if this same mouse is then put into contact with a "normal" microbiota and fed a balanced diet, its condition improves. These observations suggest that while dysbiosis may be responsible for some characteristics of these disorders, it can be corrected. At the Institut Pasteur, Philippe Sansonetti's team in the Molecular Microbial Pathogenesis Unit has demonstrated, again in mice, that a high-fat diet disrupts the balance of the gut microbiota and changes its position in the intestine, which in turn alters the workings of the intestinal wall—but that this process is reversible.

In humans, dysbiosis has been demonstrated in several conditions involving inflammation or immune dysfunction. One example is the autoimmune disease lupus. Gérard Eberl's team has elucidated the mechanism used by the microbiota to act on the immune cells responsible for triggering allergies. Links have also been discovered between dysbiosis and the severity of Parkinson's disease, for example, or between dysbiosis and breast cancer. More surprising still, dysbiosis has been observed in children with autism. Studies in mice have also shown that the gut microbiota is involved in

An autoimmune disease is one in which the patient's immune system attacks his or her own healthy cells.

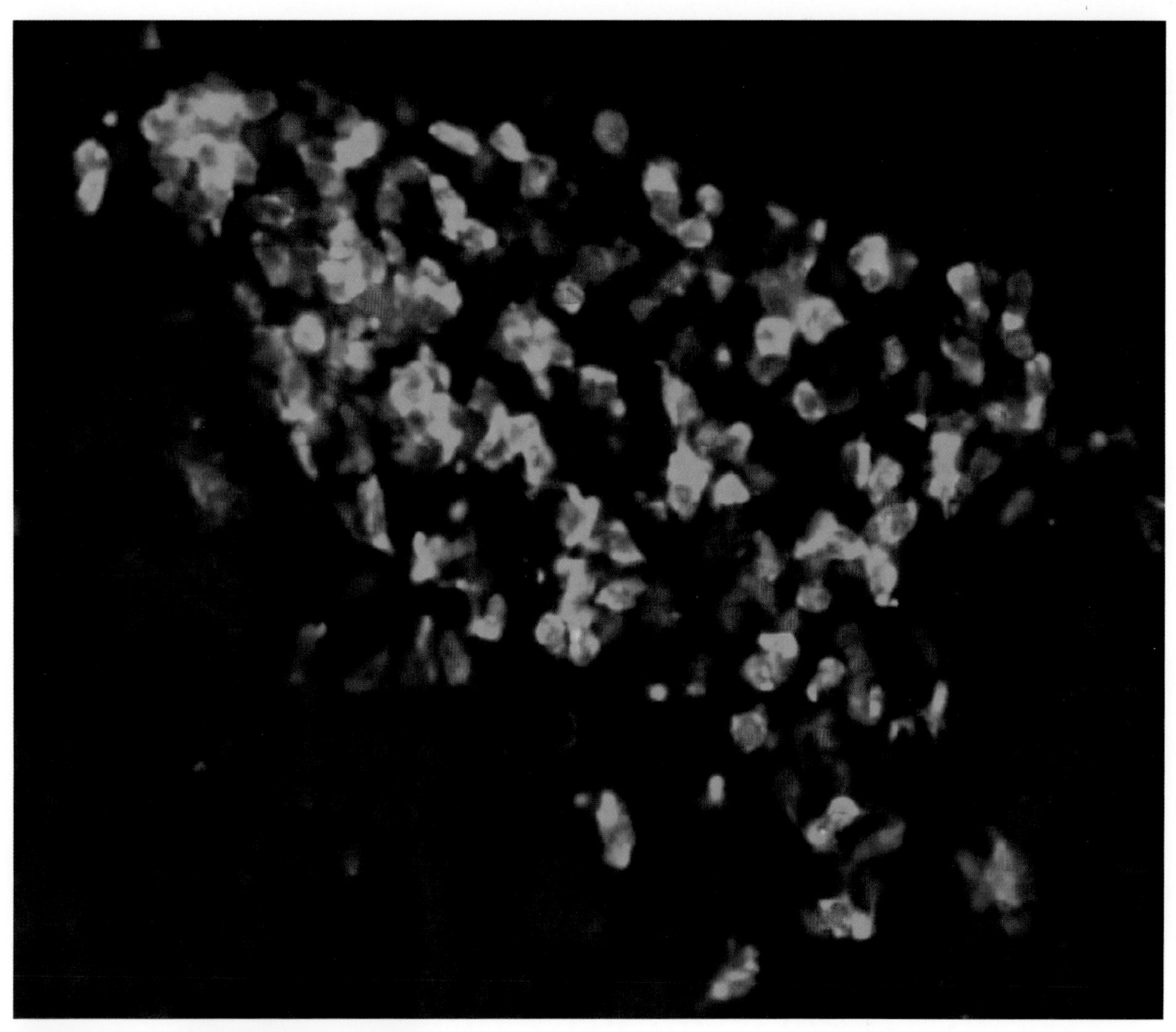

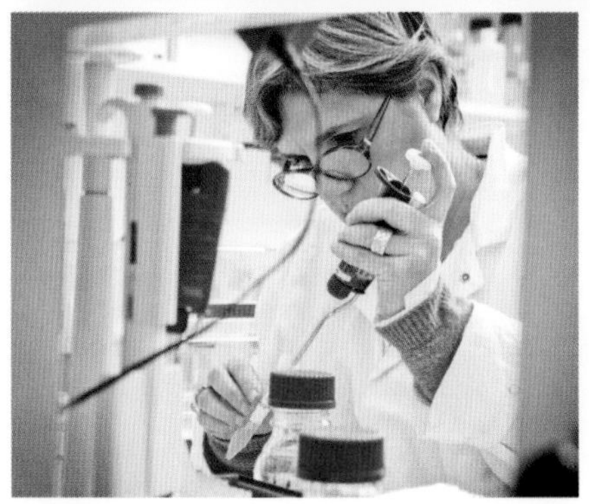

susceptibility to stress. To shed more light on these links, a Major Federating Program entitled "Microbiota and Brain" has been launched at international level, with 17 teams specializing in neuroscience, microbiology and immunology—including several teams from the Institut Pasteur—investigating the influence of the microbiota on the nervous system.

Top: A cluster of type 3 cells (in green) in a mouse colon. These cells are induced by the microbiota and block type 2 allergic reactions.

Bottom: Photo taken in the Bacteria-Cell Interactions Unit, which focuses much of its research on *Listeria*, the listeriosis bacteria. The team recently discovered that these bacteria secrete a toxin that damages the gut microbiota and promotes infection.

The microbiota is sensitive to our lifestyle

More generally, some scientists and epidemiologists see a clear link between changing lifestyles over the past 50 years (for example the increase in cesarean births and antibiotic use and the reduction in fiber intake), the consequences on the body's symbiosis with the gut microbiota, and the rise in some diseases involving immune dysfunction or chronic inflammation. The delicate balance between humans and their microbiota that has developed over millions of years of evolution could be undone by our modern lifestyles—and our health is suffering as a result.

Treating the microbiota therefore seems to be an increasingly obvious therapeutic strategy. By adjusting our intake of fiber or pre-/probiotics, scientists hope to be able to influence these disorders. Several avenues have been explored, but changing the balance of a "superorganism" remains a difficult task.

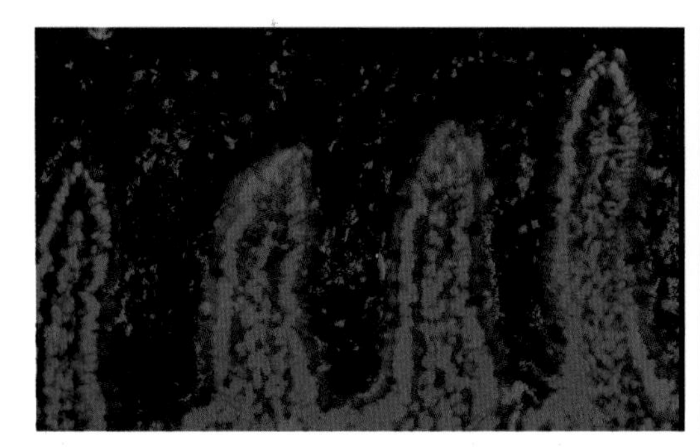

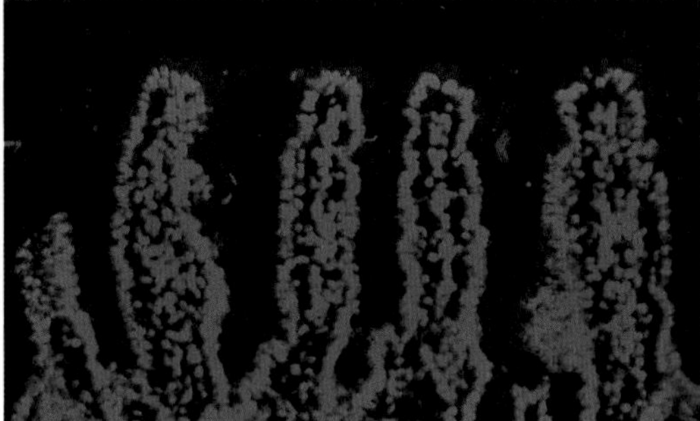

> "We are trying to improve our understanding of how we coexist with our microbiota and what role our immune system plays, not just as a system of defense but as a system of coexistence that enables us to achieve that balance."
>
> Gérard Eberl, Head of the Microenvironment and Immunity Unit at the Institut Pasteur

Localization of bacteria in the ileum of mice given a normal diet (left-hand image) and a high-fat diet (right-hand image).

THE OTHER MICROBIOTAS IN OUR BODY

The gut microbiota is our largest microbiota in terms of size and diversity and also the number of interactions it has with our body. But it is not the only microbiota we have: every surface of our body that is in regular contact with the external environment has its own microbiota. The skin and all of our mucous membranes (the nose, mouth, lungs, vagina, penis, etc.) also have their own ecosystems which perform functions that are vital for our health. Each actively participates in its own way in the efficacy of our immune system.

The skin microbiota covers an area of approximately 1.50 to $2m^2$ in adults. Its composition varies from one region of the body to the next (hands, armpits, face, etc.). We know that it is involved in common skin conditions such as acne and psoriasis, and in the chronic nature of some wounds. Research has also demonstrated that it plays a major role in our olfactory signature, in other words the natural odor of our skin, and that its balance is affected by the perfumes, creams and other products we use. The cosmetics industry is becoming increasingly interested in this field, with companies even looking into designing special fragrances for the different skin microbiotas in our body and in the population.

We have known for several years that the **vaginal and penile microbiotas** play an important role in protecting us against some infections, especially sexually transmitted infections (STIs). Studies have shown, for example, that an abnormal vaginal or penile microbiota affects the risk of acquiring HIV. In women, the vaginal microbiota produces lactic acid, which helps protect against some local infections that cause vaginitis (a local inflammation often associated with mycosis).

The lung microbiota is the least well known of our microbiotas and has so far given rise to scant research. In January 2017, a French-Belgian study demonstrated for the first time that some bacteria which are naturally present in the lung microbiota may be associated with beneficial or harmful effects on the symptoms of asthma. This discovery opens up promising new therapeutic possibilities.

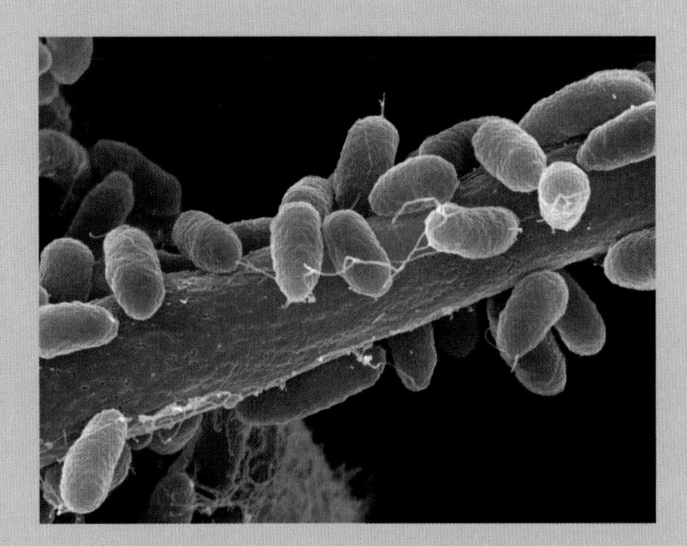

Interaction between *Aspergillus fumigatus* and *Pseudomonas aeruginosa* observed using scanning electron microscopy.

Unlocking the Brain's Mysteries

THE BRAIN IN FIGURES

80 TO 100 BILLION
neurons in our brain...

10,000
connections (synapses) made by each neuron with other neurons on average

20%
of our body's energy is consumed by the brain while it accounts for only 2% of the body's weight

The human brain contains some 86 billion neurons, or nerve cells—the basic functional units of the nervous system—and just as many glial cells that are needed for them to work properly. In comparison, the processors in today's most sophisticated computers have several billion transistors, their basic electronic components. While the computing capacity of modern computers outpaces that of the human brain, the brain uses a different approach to achieve a comparable result. In the brain there are no fixed electronic circuits; neurons are organized into dynamic networks, with multiple connections that are constantly changing in space and time. On average, each neuron establishes 10,000 contact points (synapses) with other neurons. This ability to create and adapt networks throughout life, known as plasticity, offers enormous potential for adaptation to changes in the environment or the body. It provides a remarkable degree of flexibility in behavioral terms, especially as a result of interactions between various areas of the brain specializing in language, object recognition, visual information processing and motor control. And by constantly recreating new connections, the brain has the ability, to some extent at least, to compensate for the harmful consequences of some types of damage, such as brain lesions resulting from a stroke.

Opposite: TNT (tunneling nanotube) connecting two neuronal CAD cells.

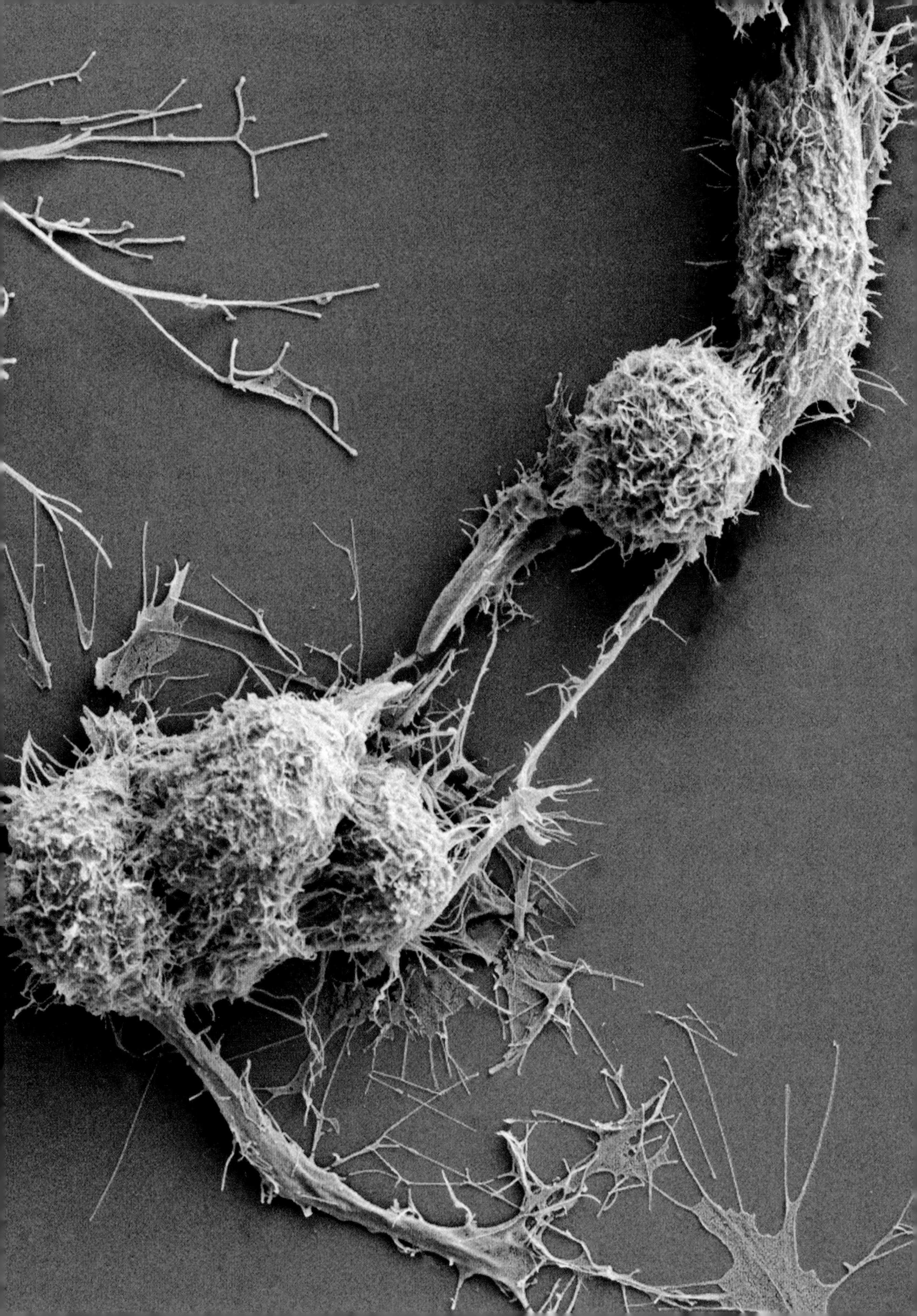

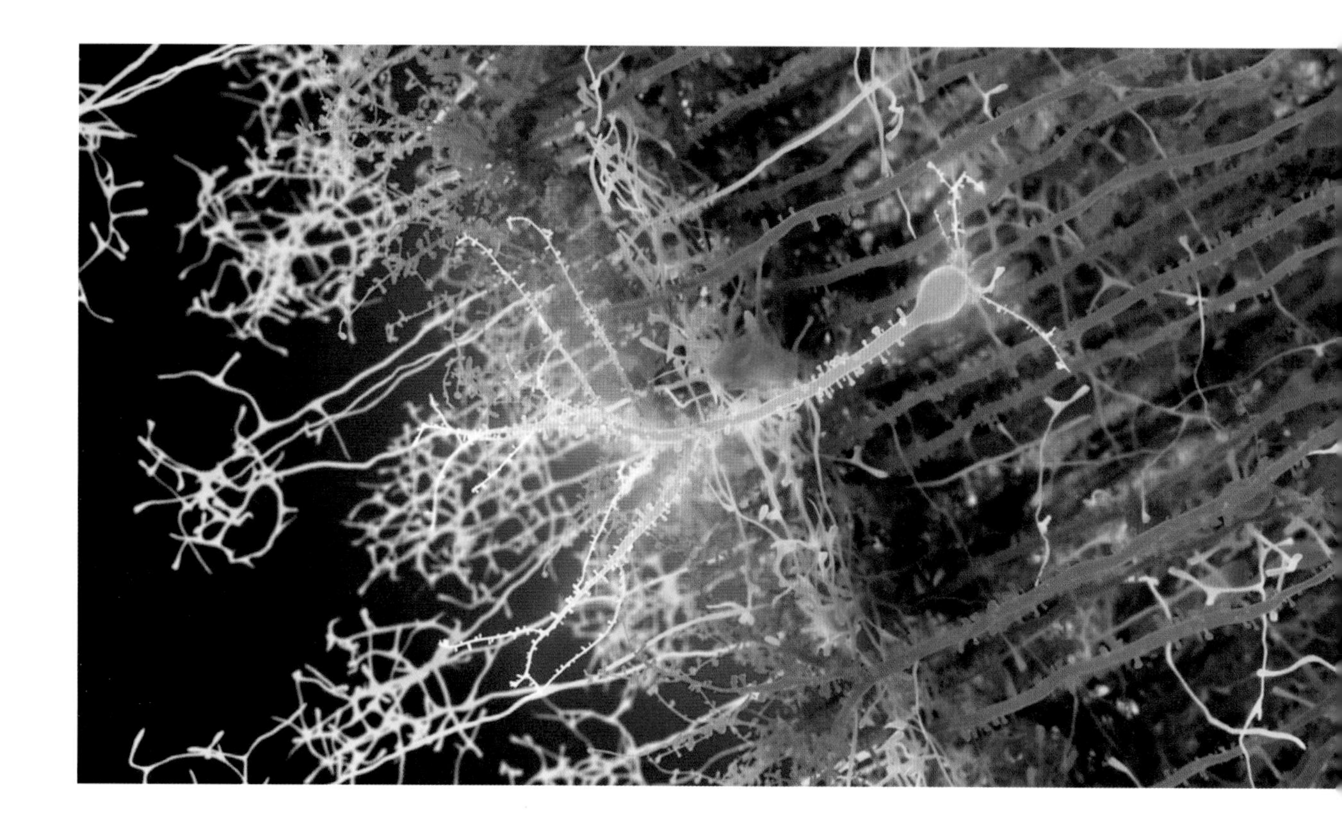

This dynamic view of the brain has been developed in recent decades on the basis of two main avenues of research. Firstly, progress in cellular and molecular biology has given us a more precise understanding of neurons: their morphological and neurochemical identity, the number of synapses they form and the way they function at molecular level. Scientists are now able to decipher the communication codes between neurons, identify the signals that activate or inhibit them, and determine their internal molecular mechanisms. Secondly, there have been significant developments in brain scanning techniques, for example functional magnetic resonance imaging (fMRI), a method in use since the 1990s that is based on measuring the brain's oxygen consumption. fMRI makes use of the fact that neuronal activity causes an upsurge in oxygenated blood. By measuring these changes in blood flow, it provides a real-time image of the regions that are activated when we carry out a given task or feel a certain emotion, or when our senses are solicited.

In light brown, in the center of the image, a new light-activated adult-born neuron. The neurons in blue are synaptic partner neurons, which connect to the new neurons. The neurons in dark brown are pre-existing neurons.

However, these two approaches cannot provide a seamless view of brain function from molecular to behavioral level. At cellular or molecular level, our observations of the brain are partial and fixed, whereas neurons communicate using extremely short electrical signals—lasting roughly a millisecond—and several chemical messengers. Although new developments in imaging instruments aim to help us understand the dynamics of the brain, they do not yet offer a high enough spatio-temporal resolution to display the unique activity of each neuron.

The challenge of neural networks

Modern neuroscience tries to link these two scales by investigating the brain at an intermediate level, namely that of neural circuits: neuron populations that have been specifically activated by a given cerebral and behavioral activity. This is a major challenge. It is at this level, for example, that memory is stored as strengthened connections, or that decision-making occurs as a result of exchanges between neurons. Elucidating the dynamics of these networks, at the level of their component units, is a way of identifying the multiple sequences of neural events that phenomena such as mental representations and decision-making are based on. For the past few years, this goal has seemed to be within our grasp. Neuroscientists now have tools and techniques which offer them an unprecedented degree of resolution to investigate the complexity of neural circuits, and they can realistically hope to understand the functional interactions between the neurons in our brain at a higher, more integrated level.

Today's powerful computers can also be used to model neural networks. Since each neuron has thousands of contacts with other neurons, considerable computational capacity is required. These models offer new working hypotheses that can be used in conjunction with experiments, which in turn refine theoretical models further in a continuous beneficial cycle.

OPTOGENETICS

New techniques such as optogenetics now enable us to view and manipulate neurons. Optogenetics, developed in the 2000s by Karl Deisseroth's team at Stanford University, uses bursts of light to control the production of electrical impulses in a targeted group of neurons, with millisecond precision. Laboratories across the world are now using this technique to specifically activate or inhibit groups of neurons, analyze their functional interactions with other neuronal assemblies, and identify how they contribute to psychological and behavioral functions.

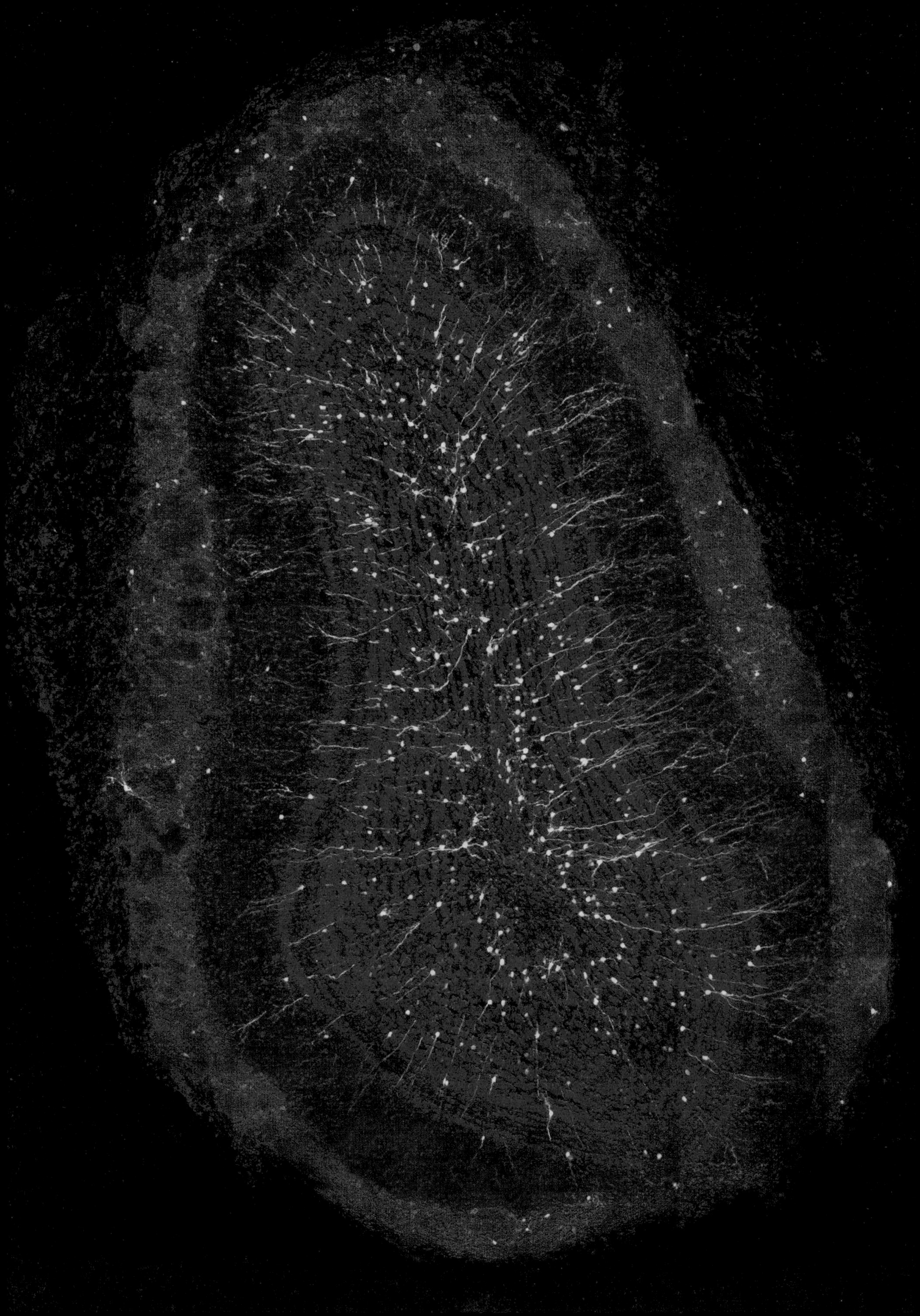

The brain's role in smell

The Institut Pasteur's Department of Neuroscience investigates how the brain works at all levels, from genes to behavior. Several teams are exploring neural networks to try to shed light on the role of neurons or neural circuits in behavioral tasks.

Gabriel Lepousez, Gilles Gheusi and their colleagues in the Perception and Memory Unit, directed by Pierre-Marie Lledo, are combining optogenetics and imaging techniques to try to understand how the perception of an odor can give rise to a specific emotion. Our sense of smell is constantly stimulated by different scents that trigger a variety of reactions: a smell of chocolate or freshly baked bread makes us hungry, while a smell of mold elicits disgust. These emotional reactions are crucial for survival, helping us to avoid toxic products and alerting us to potential danger. The olfactory bulb is a key brain region involved in the perception and memory of smells. This small area is connected to the olfactory sensory neurons that line the nasal cavity and detect the arrival of scent molecules. By means of reciprocal connections with other regions of the brain, it serves as the first relay station in the olfactory system, determining our behavioral reactions to a given smell. However, little is known about the circuits and mechanisms that govern these functions and their interactions. By activating or inhibiting specific neurons or neural circuits, the team hopes to determine how the memory of a smell is coded in the networks of the olfactory system and how these smell associations influence our behavior. The team is particularly focusing on the role played by the 30,000 new, or adult-born, neurons that are added to the olfactory bulb in mice every day.

ADAPTATION OF THE OLFACTORY SYSTEM

Pierre-Marie Lledo and his colleagues are also exploring the genesis and dynamics of new neurons in the olfactory bulb to shed light on how the olfactory system constantly adapts to environmental changes. In 2016, working with scientists from Johns Hopkins University in Baltimore, they monitored new neurons in the olfactory bulb of mice using a fluorescent protein produced only in these neurons. The scientists observed that their connections with other neurons remained highly dynamic throughout their life. Using computer-based models, they also showed that this dynamism enables the synaptic network to adjust quickly and effectively to sensory changes in the environment. The next stage is to find out how new neurons can be diverted to regions that are damaged in neurodegenerative conditions, such as Huntington's chorea or Parkinson's disease, so that they can restore affected nerve circuits.

Opposite: Dopaminergic and newborn neurons in an adult murine olfactory bulb. New active neurons integrate the olfactory bulb circuit in mice daily, even in adulthood.

For many years it was thought that new neurons were only developed in the period from embryogenesis to puberty. But in the late 20th century, it was conclusively proven that in adults, new neurons are created in two brain regions: the hippocampus, where our memories are formed and our emotions are controlled, and the subventricular zone (SVZ), which, in humans, supplies neurons to another region that deals with motor control and reward systems. In rodents, approximately half of the neurons generated in this small area are eliminated, but the other half migrate to the olfactory bulb, where they are involved in developing olfactory memory. Scientists at the Institut Pasteur are investigating how these new neurons are involved in the processes governing sensory perception, memory and cognitive flexibility. In 2014, they showed that new neurons begin to mature in the deepest region of the olfactory system, the region that is closest to the cerebral cortex. The new neurons in the olfactory bulb are connected to multiple other regions in the brain, providing the conditions needed to associate a smell with its context. Neuroscientists have observed that new neurons create a greater number of strong connections with other brain regions if, for example, an individual associates a specific smell with receiving a reward. But much still remains to be discovered about how associations between smells, emotions and memory are developed in individuals, and how they are impaired in those with behavioral disorders related to conditions such as depression and schizophrenia.

Photo taken in the Perception and Memory Unit in 2011.

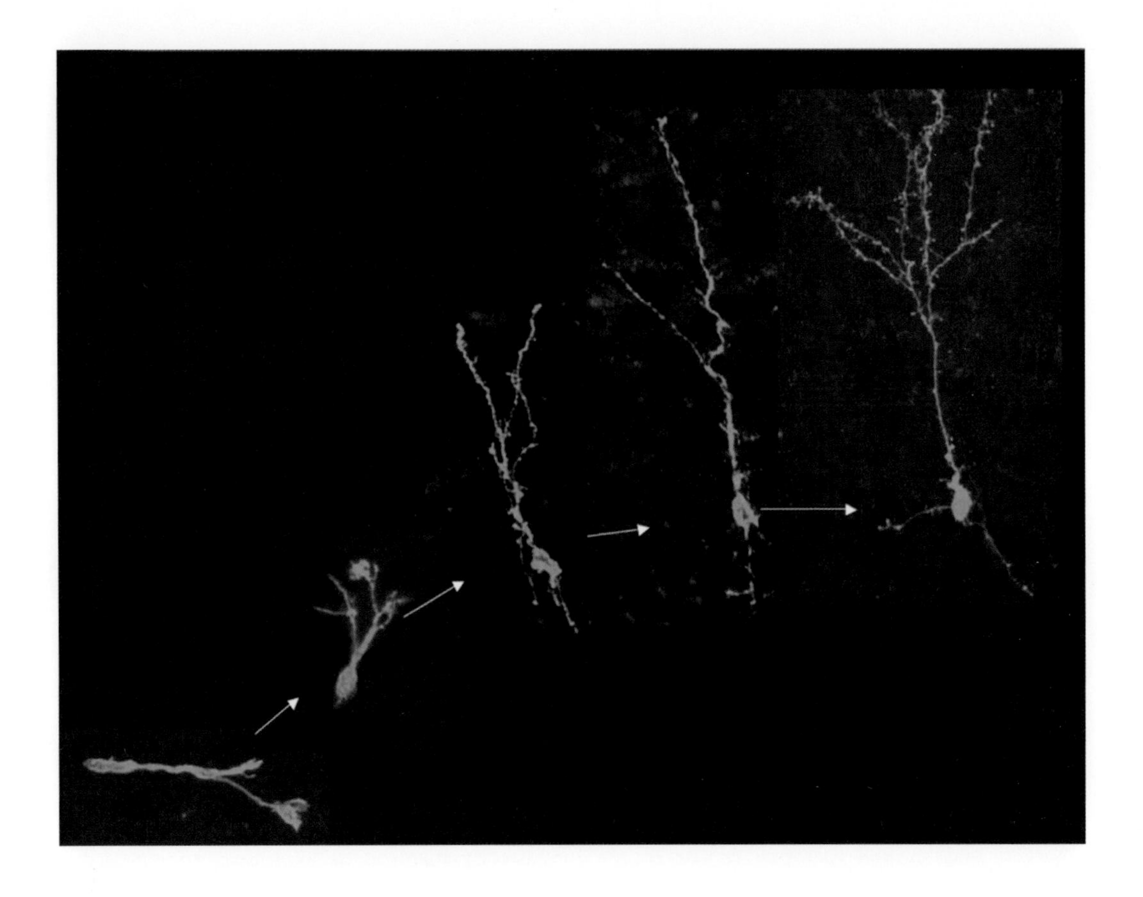

“We are looking for methods and strategies that enable us to target, modulate or even specifically repair non-functioning circuits, without disrupting the entire brain.”

Gabriel Lepousez, scientist in the Perception and Memory Unit at the Institut Pasteur

Formation of new neurons in the mouse olfactory bulb.

From nicotine to schizophrenia

Even something as seemingly simple as a neuronal protein can be investigated at several levels. This is the approach adopted at the Institut Pasteur by the teams led by Uwe Maskos, from the Integrative Neurobiology of Cholinergic Systems Unit, and Pierre-Jean Corringer, from the Channel Receptors Unit. The focus of their research is a protein located in the neuron membrane which is specifically associated with a chemical messenger found in neurons, known as acetylcholine. When acetylcholine molecules come near the protein, they bind to it. The protein then forms a pore in the membrane through which ions can enter the neuron and activate it. This mechanism is involved in several processes such as control over voluntary movements, memory, attention, sleep, pain and anxiety. The protein is known as an acetylcholine receptor or a nicotinic receptor, since nicotine also binds to it with high affinity, triggering the same neuron activation mechanism. This high-affinity binding gives rise to nicotine addiction by activating pleasure-inducing neural circuits.

Research into nicotinic receptors goes back a long way at the Institut Pasteur: in the 1960s, Jean-Pierre Changeux led the first neuroscience research with his investigations into these proteins (see page 21). The teams directed by Uwe Maskos and Pierre-Jean Corringer are continuing this work by examining all aspects of the receptor, from its molecular structure to its role in disorders such as schizophrenia and addiction, as well as in neurodegenerative diseases such as Alzheimer's. In 2016, they determined its three-dimensional structure and showed that part of the receptor is directly involved in Alzheimer's disease. In mice that do not produce that part of the receptor, the disease does not result in memory loss. The scientists are now trying to find therapeutic applications by identifying a molecule that blocks this part of the receptor and counters the memory loss experienced by Alzheimer's patients.

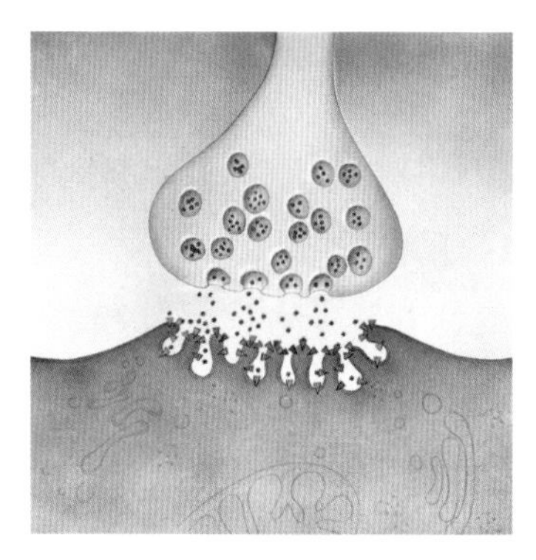

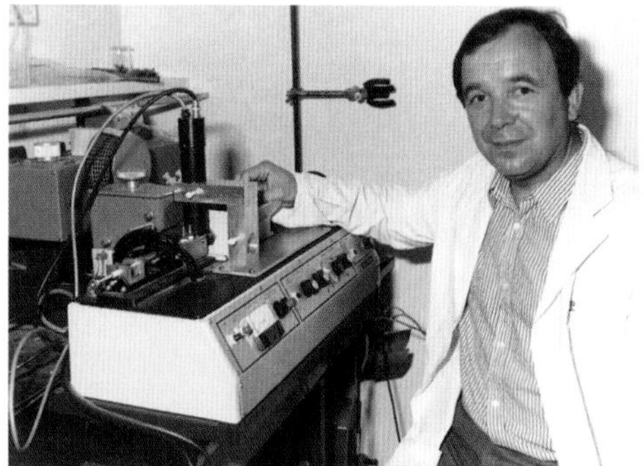

Above left: Diagram of a synapse. The acetylcholine—or nicotine—released in the synaptic cleft binds to receptors carried by the cell (green triangles), resulting in the opening of the ionic channel. A chemical signal (acetylcholine binding) is therefore converted into an electrical signal (flow of ions through the channel).

Above right: Professor Jean-Pierre Changeux, around 1980.

Opposite: Acetylcholine receptors isolated and purified by Jean-Pierre Changeux at the Institut Pasteur in 1970.

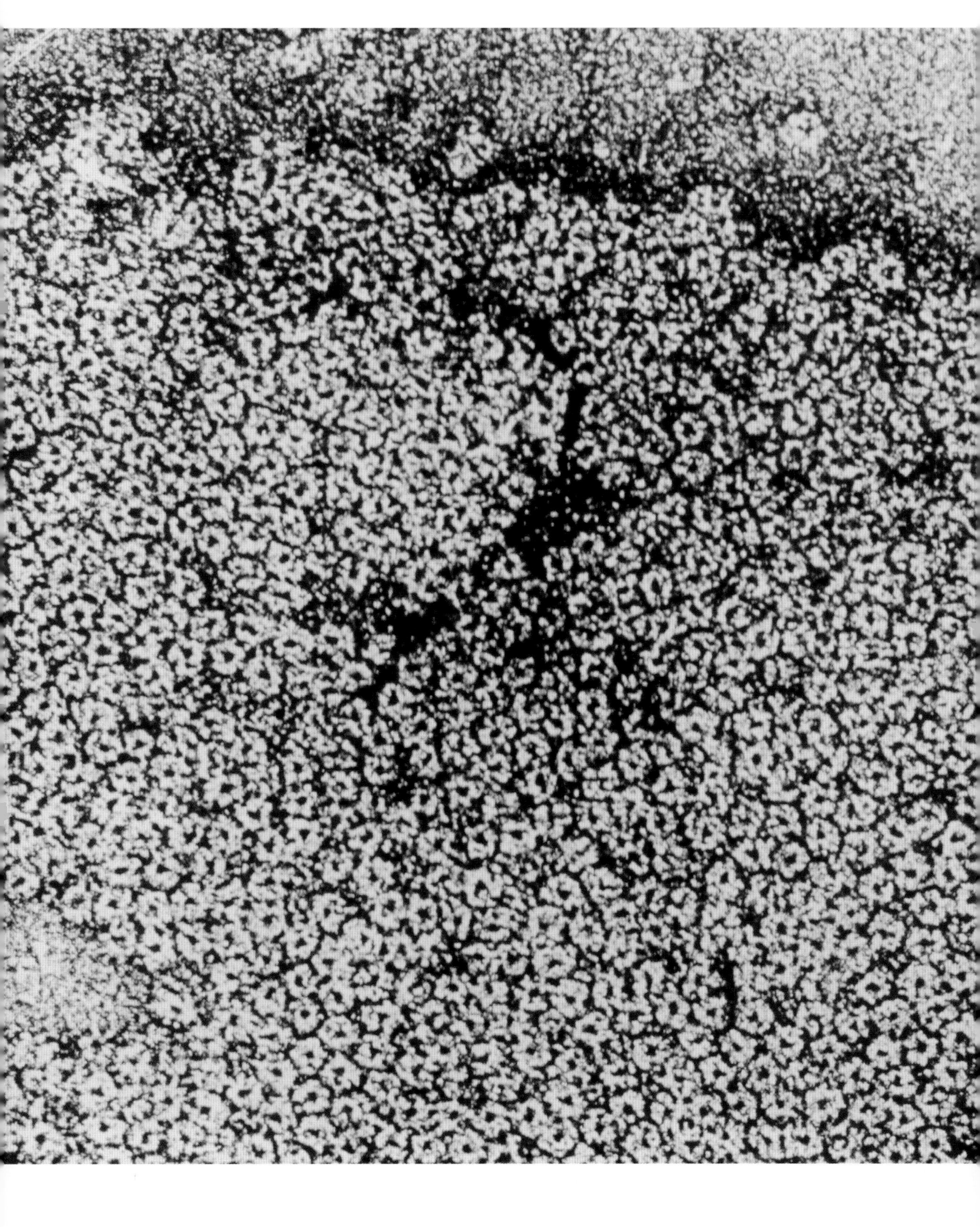

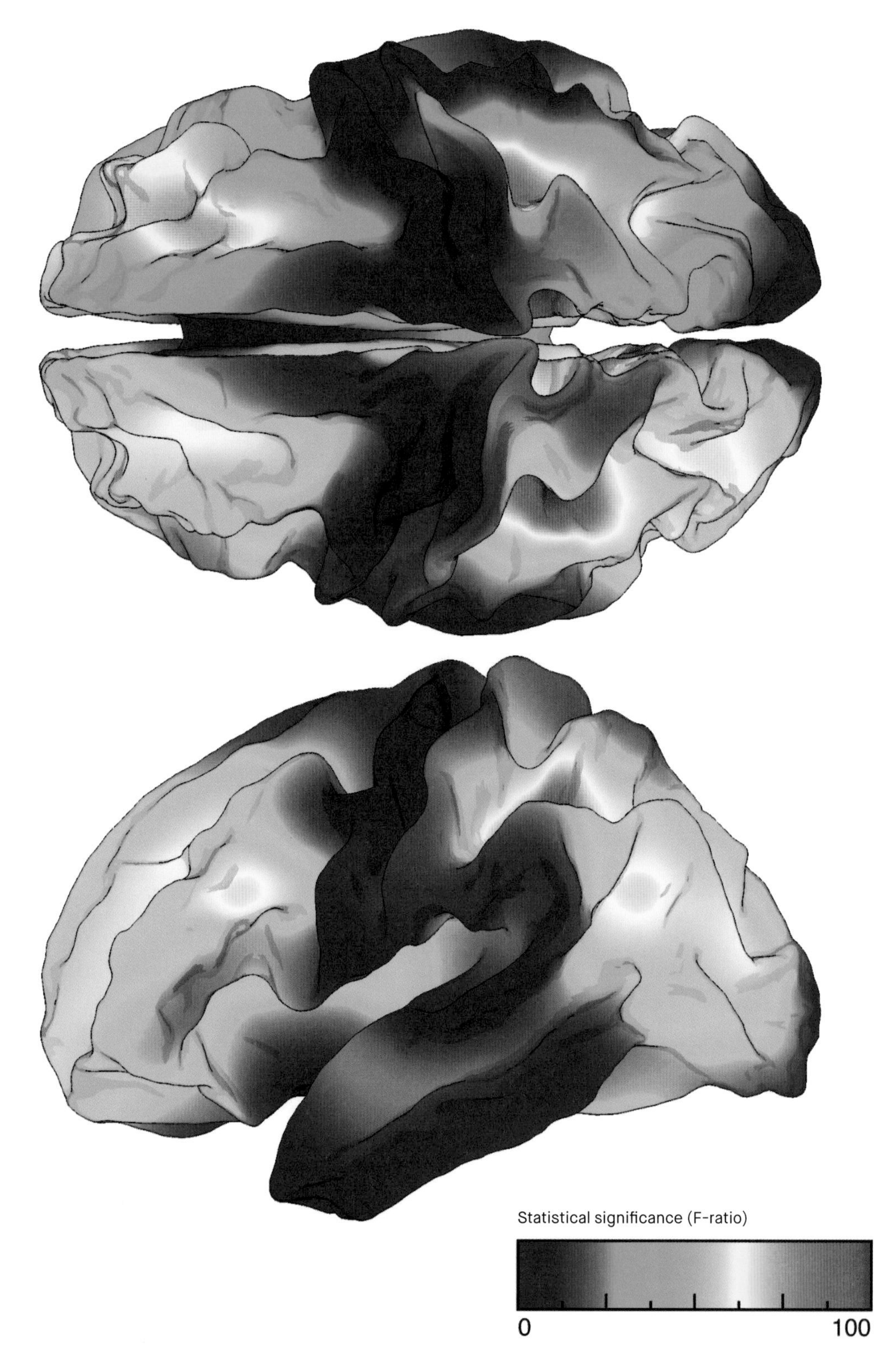

Surface map showing how cortical thickness changes with age during adolescence (Statistical significance / F-ratio).

“This approach follows directly on from the strategy introduced by Jean-Pierre Changeux at the Institut Pasteur. In 1983, he wrote a book* in which he set out the idea of linking the molecular level with more complex cerebral functions such as language, reason and conscience.”

Gilles Gheusi, scientist in the Perception and Memory Unit at the Institut Pasteur

* *L'Homme neuronal,* Jean-Pierre Changeux, Fayard, 1983.

In 2017, with scientists from the CNRS, Inserm and the École Normale Supérieure in Paris, Uwe Maskos' team also demonstrated why schizophrenics frequently use smoking to overcome the lethargy and lack of motivation caused by their treatment. These individuals often have a mutation in their nicotinic receptor. Neuroscientists introduced this mutation in mice and observed that their intermediate neurons (neurons that create links between neural networks) were less active, just as in schizophrenic patients. But when the mutant mice were given nicotine, it bound to the nicotinic receptors in these neurons and their activity levels returned to normal. The impact of nicotine suggests that a similar molecule that does not have the same harmful effects could improve the quality of life of schizophrenic patients.

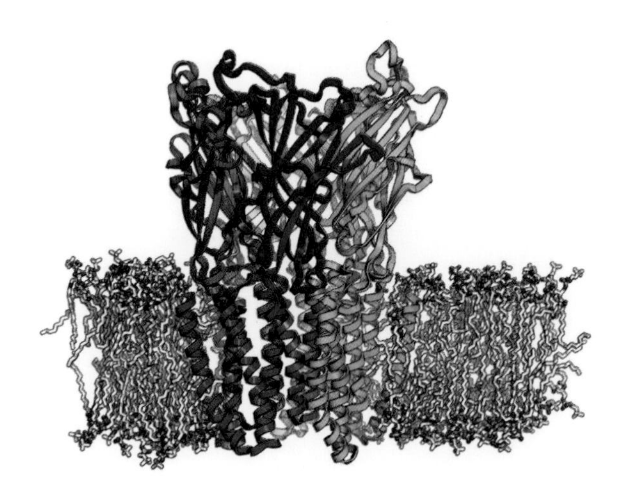

The 3D structure of a bacterial protein near the human nicotinic receptor.

The brain as part of the body

It has become increasingly clear that we cannot separate the workings of the brain from its environment, whether we are trying to understand the evocative power of scent molecules or the wide-ranging action of nicotine. And it goes without saying that the brain's primary environment is the body that houses it, which is why experts in neuroscience—a discipline closely related to psychology—continue to draw on other disciplines such as microbiology, immunology and virology to further bolster the integrative dimension of their approach to biological systems. The aim is to understand what links the brain to the rest of the body and ensures that it functions properly. For example, how do the microbes in our intestines—the gut microbiota—act on the brain? How does the immune system interact with the brain? Recognizing the same themes that first inspired its pioneering scientists, the Institut Pasteur decided to look more closely into these questions.

Together with other scientists on campus, Pierre-Marie Lledo launched a research program on the microbiota and the brain. The Institut Pasteur's neuroscientists are trying to shed light on how chronic stress, depression, anxiety and mood swings are linked with the gut microbiota. In 2017, they demonstrated that the microbiota influences processes that modulate the brain's activity by releasing molecules into the gut. These molecules are absorbed by the intestinal wall and pass into the bloodstream, which transports them to the blood-brain barrier. How do these molecules act on the brain? Do they cross the barrier? The scientists have so far only scratched the surface of this topic. If the gut microbiota is one of the many factors that govern our behavior—as a significant body of research now suggests—it opens up prospects for entirely new therapeutic strategies and avenues for exploration in terms of medical applications.

There is also a three-way relationship between the intestinal microbiota, the immune system and the brain whose mechanisms are yet to be explored. The microbiota influences the activity of the immune system, which relays the information directly to the brain, releasing small molecules known as cytokines into the bloodstream. Moreover, a German study demonstrated in 2015 that the immune cells in the brain—microglial cells—are constantly "listening" to the compounds released by the microbiota. Together with the Microenvironment and Immunity Unit directed by Gérard Eberl at the Institut Pasteur, Pierre-Marie Lledo's team is investigating the mechanisms that underpin this molecular language.

Opposite: Cells from the small intestine of mice (in green) and bacteria from the microbiota (in red).

“In neurological conditions such as Alzheimer’s and Parkinson’s, inflammation is a major factor in neuron destruction, involving the brain’s immune cells. If the microbiota controls these cells remotely, we want to identify exactly how it is involved in the neurodegenerative process.”

Gabriel Lepousez

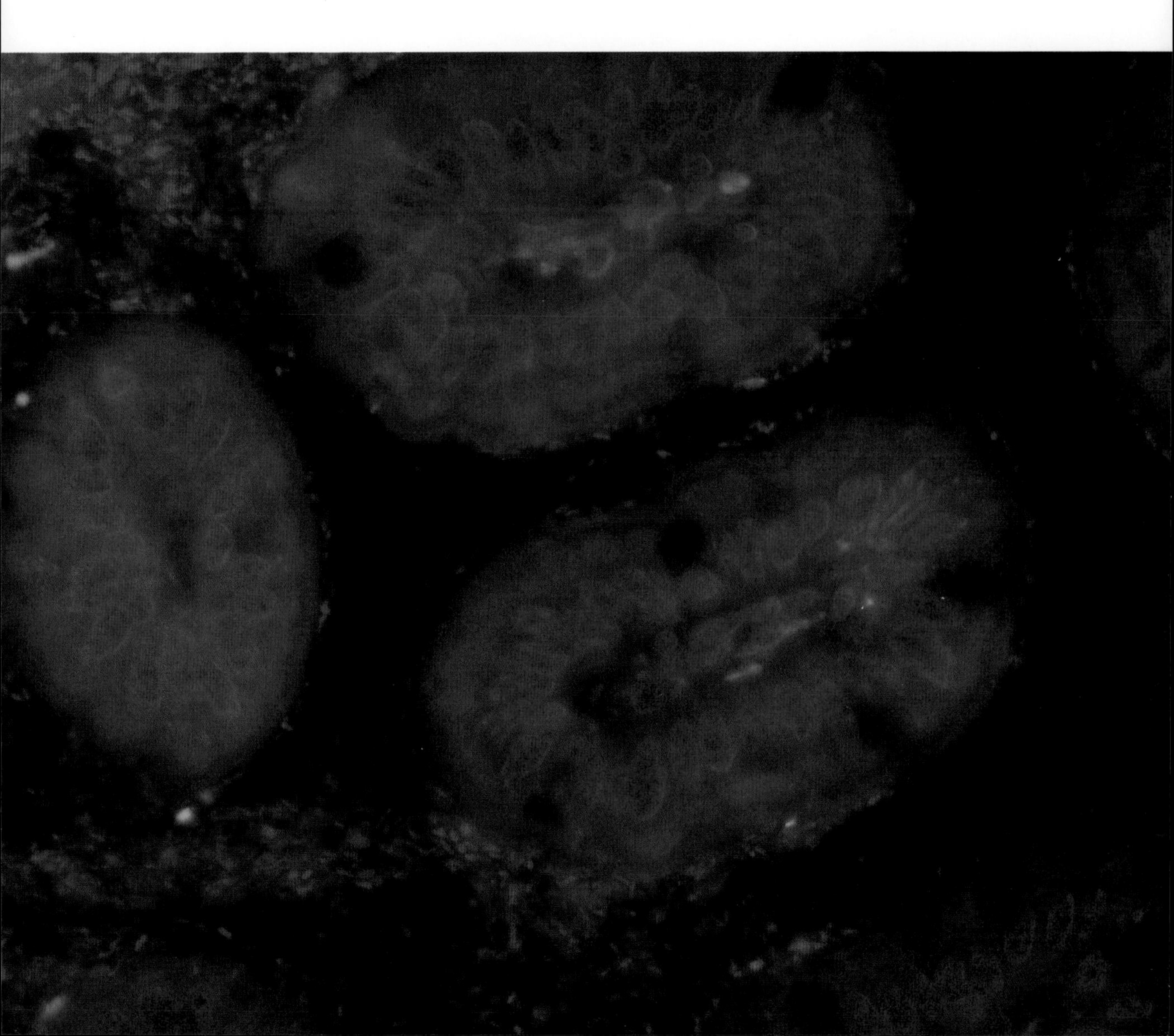

AN ELIXIR OF YOUTH

"I believe there is an 'elixir of youth' waiting to be discovered!" explains Lida Katsimpardi, scientist in the Perception and Memory Unit at the Institut Pasteur. It is now clear that investigating the brain as an isolated organ shut away in its skull results in an overly simplistic, incomplete view of the way it works. In 2014, Lida Katsimpardi, then at Harvard University, and her colleagues discovered that the body produces hormonal compounds that play a considerable role in governing the processes related to brain rejuvenation and aging. When the scientists gave blood taken from young mice to a group of older mice, it increased the production of new neurons in these older mice, leading to vascular remodeling in the brain. The olfactory discrimination capabilities in the older mice improved and they did not exhibit the memory loss traditionally associated with aging. The biologists successfully isolated a blood factor known as GDF11, which reproduced virtually the same effects on its own. Lida Katsimpardi, who has since joined the Institut Pasteur, is now looking for other rejuvenating molecules in the blood.

Lida Katsimpardi from the Perception and Memory Unit. Imaging using confocal microscopy to examine neurogenesis in the hippocampus of old mice.

> “Mental representations do not just reflect brain activity. They are based on both brain activity and the activity of other structures in the body that enable us to perceive and act in accordance with our environment.”
>
> Gilles Gheusi

The vagus nerve is a key player in these interactions, and its role is only just beginning to be understood. This nerve, which runs from the brain to the digestive system via the heart and the lungs, has a twofold purpose. First, it acts virtually independently to orchestrate vital processes for the body such as the contraction of muscle fibers in the heart, lungs and intestines, and the secretion of digestive enzymes. In this role it has an effector function, relaying a series of cerebral activities to an effector such as a gland or a muscle. Second, it serves as a sort of probe, a sensory branch of the brain that provides it with information about what is happening in the body. The vagal nerve endings have several receptors for cytokines and for bacterial compounds. In other words, this receptor function enables the vagus nerve to inform the brain about the working condition of the immune system and activities in the gut microbiota. It therefore merits investigation alongside the other sensory systems.

Navigating the jungle of human behavior

The body is not the only environment that has an impact on the brain. Adopting a more comprehensive approach to the brain and trying to understand why a given situation results in a given behavior also means looking at the interactions between individuals and the world around them and their potential scope for activity. That is why scientists are now focusing on investigating the whole gamut of human and animal behavior. After the genome (the genetic makeup) and other comprehensive "ome" data that we can now collect about humans as a result of recent progress in bioinformatics (see page 167), they are turning their attention to the "ethome". The idea is to increase the volume of data available about the incredibly diverse repertoire of human behavior.

In the past, behavior was always described via the lens of human perspective, which skewed the analysis. For example, we only recorded what we were able to recognize and express. Progress in computing techniques means that we can now overcome this obstacle. Today's sophisticated machines can identify repeated behavior in experiments (an arm movement, a posture, etc.) which would previously have been impossible for humans to detect, and then isolate this particular behavior. For example, eating behavior is not just about the body's search for food. It also involves motor, sensory, motivational, relational and memory-related aspects. Based on hypotheses and independent tests carried out using existing databases, a computer with an artificial intelligence system can help identify and separate all these components.

The task is a huge one, and neuroscientists are approaching it from all angles. Teams are gathering data or developing new data-collection tools.

These tools can generate considerable volumes of personal health data and could have applications in medical research, especially in behavioral science. With the development of effective algorithms, the big data revolution could enable us to predict risks and personalize prevention and treatment by improving health care monitoring and services. Used advisedly, connected devices represent an opportunity to develop a medical approach that is no longer "post-traumatic" but preventive, personalized, to some extent predictive, and also less costly.

APPLICATIONS FOR MENTAL HEALTH

At the Institut Pasteur, Pierre-Marie Lledo is working in conjunction with hospital psychiatrists (from Sainte-Anne in Paris and Henri-Mondor in Créteil) to develop applications that can provide a more precise behavioral analysis of patients by recording various parameters in real time. We know that mobile applications are transforming human behavior in the digital age—estimates suggest that more than 4 billion smartphones and tablets will be in use worldwide in 2018. Mobile applications are already available to count how many steps we take and measure our heart rate, but connected devices could also have an impact in the area of mental health. Mental disorders are widespread and take many different forms, including anxiety, depression, addiction, schizophrenia and obsessive compulsive disorder, and they represent a considerable burden on healthcare costs (€800 billion a year in Europe alone). Connected devices could be used to monitor psychiatric conditions more closely and enable patients to take an active role in their own treatment, or to help them identify risk behaviors and propose alternatives that represent less of a threat to their health.

Fabrice Hyber, Digestion de données (data digestion), 2010. Artistic production in the Organoïde project: a database of images created by artists to represent the Institut Pasteur's scientific research.

At the same time, major projects such as the Human Brain Project in Europe and the Brain Initiative in the United States are generating a corpus of data with the aim of producing a model of the brain and unraveling links between brain function and behavior. The Seattle-based Allen Institute for Brain Science, launched in 2003 by Microsoft co-founder Paul Allen, goes even further. It offers an open website with a wide range of data about the brain, such as a map of the genes translated into proteins in the various brain regions, the responses of a given neuron to a given stimulus, and the network of neural connections in a mouse brain. Scientists can use this vast database to compare their results with existing data and add their own data in turn.

When they have identified a behavioral trend, they can try to link this trend with a neural circuit, examining whether it is likely to give rise to functional deficits during a disease and whether techniques such as optogenetics can be used to correct it in preclinical trials based on an animal model of the disease. The strategy outlined by Jean-Pierre Changeux in 1983 in his book *L'Homme neuronal*—"breaking down the barriers separating neural and mental approaches and building a bridge, albeit a fragile one, from one to the other"—finally seems to be becoming a reality.

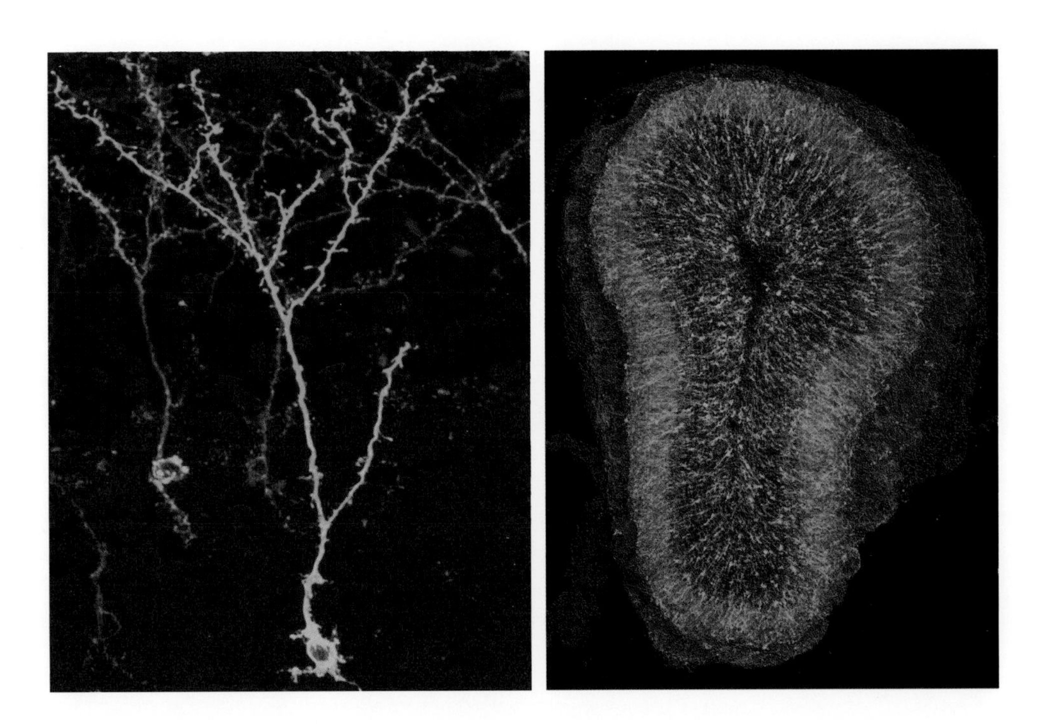

Above left: Mouse new neuron viewed under fluorescence microscope.

Above right: Section of a mouse brain observed using a fluorescence microscope. The green filaments represent new neurons in an organized network.

On the Genetic Trail
of Autism

AUTISM IN FIGURES

MORE THAN 800
genetic mutations have currently been identified in ASD patients

1 IN EVERY 100 CHILDREN
are thought to be affected by autism spectrum disorder (ASD) in developed countries

In
72%
of cases, the cause of autism is still unknown

400,000 TO 500,000
people are affected by ASD in France

Genes discovered by scientists with a direct or indirect link to ASD (via a related disorder) are involved in just
28%
of cases

For many years, autism treatment was mainly based on psychiatry. But family-based research, followed by more extensive molecular analysis, revealed that genetics, or innate factors, play a key role in what we now refer to as autism spectrum disorder (ASD). This finding completely transformed the approach to treating children with ASD.

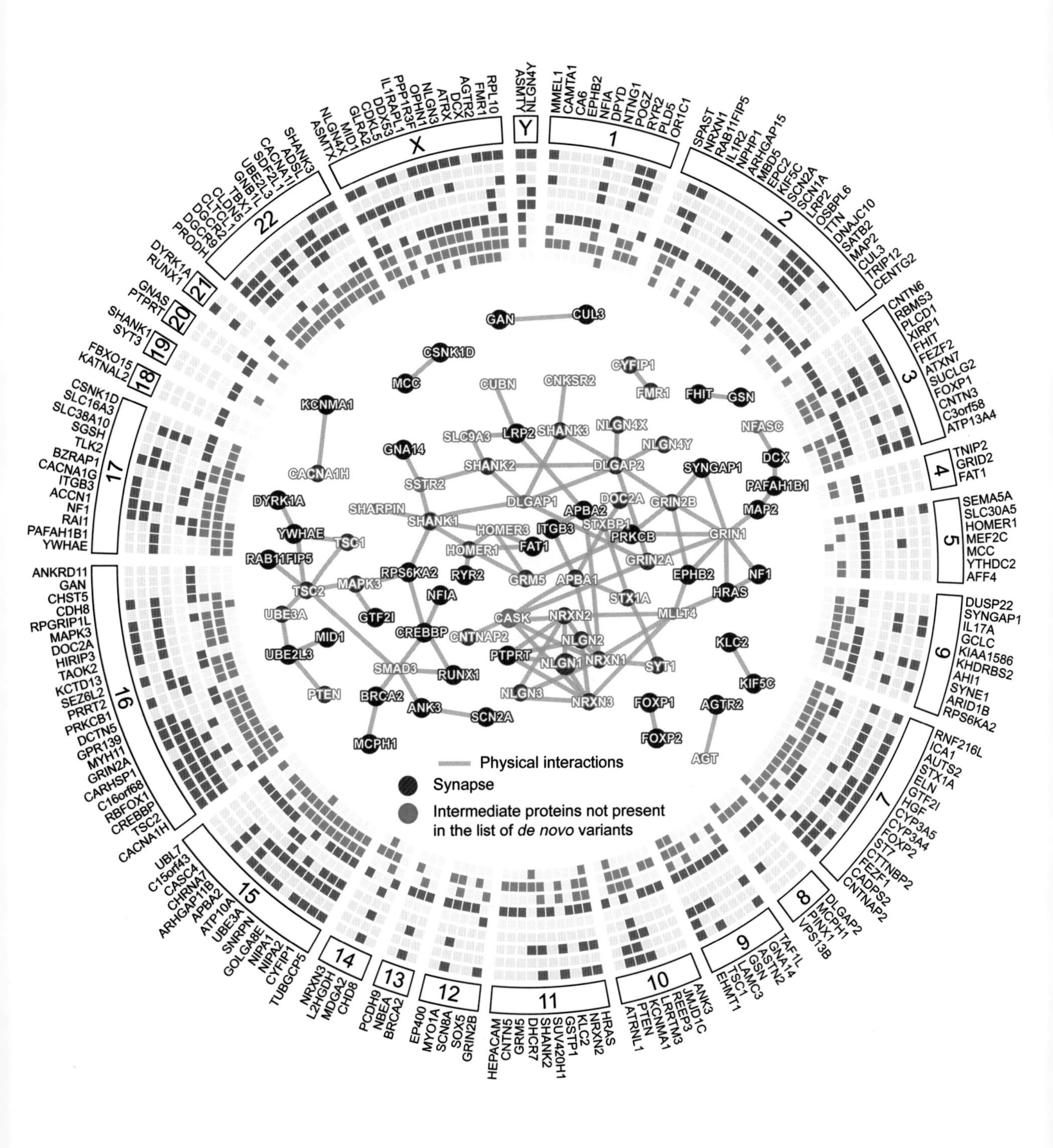

Map of autism vulnerability genes in the whole genome. Although there are more than 300 vulnerability genes, they converge towards the same biological functions, such as the establishment of contact points (synapses) between neurons. The middle of the circle shows autism vulnerability genes involved in synaptic plasticity.

Autism was first described in 1943 by Austrian-born American child psychiatrist Leo Kanner, who had been monitoring eleven children for several years. From that point on, specialists were divided over the origins of the disorder: are children born with autism or do they develop it? For many years, the theory that autism is "acquired" was widely accepted. Another American child psychiatrist, Bruno Bettelheim, suggested that mothers were largely responsible for the onset of this condition, that they were somehow "unwitting torturers" of their children. Bettelheim's theory was based on his personal experience of concentration camps during the Second World War; he drew parallels between the dejected attitude of some prisoners and the behavior of autistic children. For several years, the presumption that autism was an acquired psychiatric disorder dominated medical thinking—especially in France, where people with autism were generally referred to psychiatrists. But 60 years after Leo Kanner's description of autism, it was in France that genetic mutations were first identified in two families, each with two autistic children. This discovery was made at the Institut Pasteur by the Human Genetics and Cognitive Functions Unit directed by Thomas Bourgeron, in cooperation with psychiatrists Marion Leboyer from France and Christopher Gillberg from Sweden. Today it is widely recognized that autism has a major genetic component.

A complex series of disorders

Autism is characterized by three broad categories of symptoms, which can vary in severity between individuals: difficulties with social interaction; impaired verbal communication; and behavioral disorders, often involving a restricted repertoire of interests and activities and including repeated stereotyped movements (hand flapping, swaying, spinning, making complicated bodily movements, etc.). These signs can be accompanied by sensory hypersensitivity or hyposensitivity (increased or reduced sensitivity to sound, light, touch, etc.) and significant distress at any changes to the individual's environment. Autism patients frequently exhibit other clinical signs such as intellectual disability, hyperactivity, sleeping difficulties, eating disorders and epilepsy.

The description of autism has now been extended, and experts nowadays prefer to use the broad term "autism spectrum disorder" (ASD), which includes other forms of autism such as Asperger syndrome, where language is not affected and the patient has an average or above-average IQ.

The first signs of ASD appear in early infancy, generally before the age of three. In the vast majority of cases, they last throughout the patient's lifetime. If we consider the entire ASD spectrum, boys are four to eight times more likely to be affected than girls. Some forms of ASD are also directly associated with genetic conditions such as fragile X syndrome.

In France, it is thought that between 400,000 and 500,000 people are affected by ASD, or approximately 1 child in every 100. In the United States, the Centers for Disease Control and Prevention (CDC) estimates that about 1 in 68 children are affected by ASD. These high figures have led some experts to suggest that we are witnessing an autism epidemic in developed countries, without being able to explain the cause. But the increase in numbers can be explained to a large extent by the broadening of the diagnostic criteria and improvements in detecting autism.

A scientist in the Human Genetics and Cognitive Functions Unit. Research into the genetics of autism.

Uncovering the genetic basis

Experts now unanimously agree that the psychological traits of parents have no impact on the onset of autism; it is recognized as a condition with multiple factors and a significant genetic component. As early as the 1970s, research on twins and on families in general suggested that there may be a genetic basis for autism. At this time, research on twins showed that when a child was diagnosed with ASD, his or her twin was also affected in 92% of cases for identical (or monozygotic) twins—with exactly the same genetic makeup—and in 10% of cases for fraternal (or dizygotic) twins. In 2005 and 2009, similar research was carried out on a larger scale with 3,000 pairs of twins. ASD symptoms affected both identical twins 77 to 95% of the time and both fraternal twins 31% of the time.

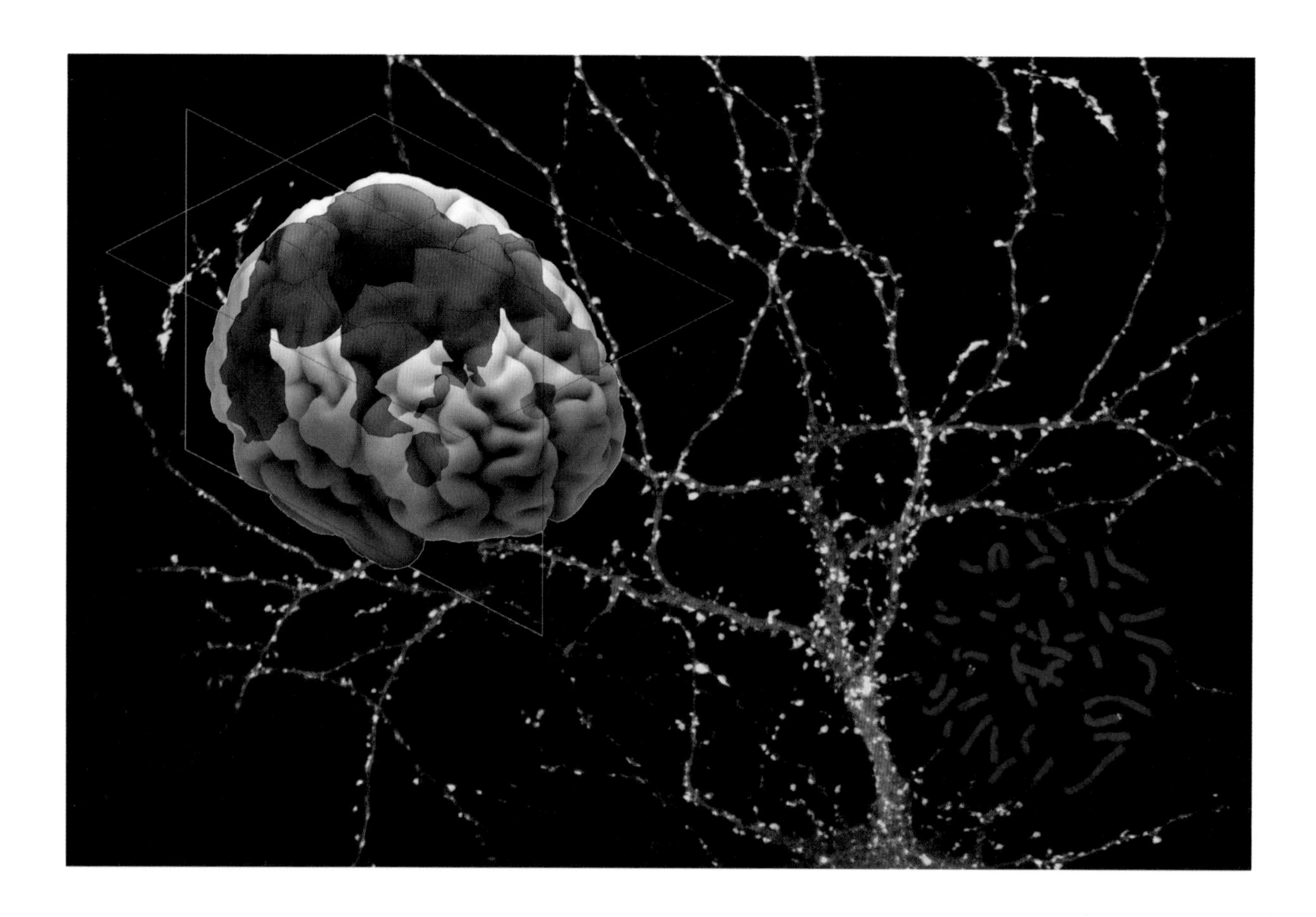

Autism is characterized by social communication difficulties and stereotypical movements. The Institut Pasteur identified the first genes associated with autism and is currently analyzing the impact of mutations on the contact points between neurons (synapses) and on brain anatomy and function.

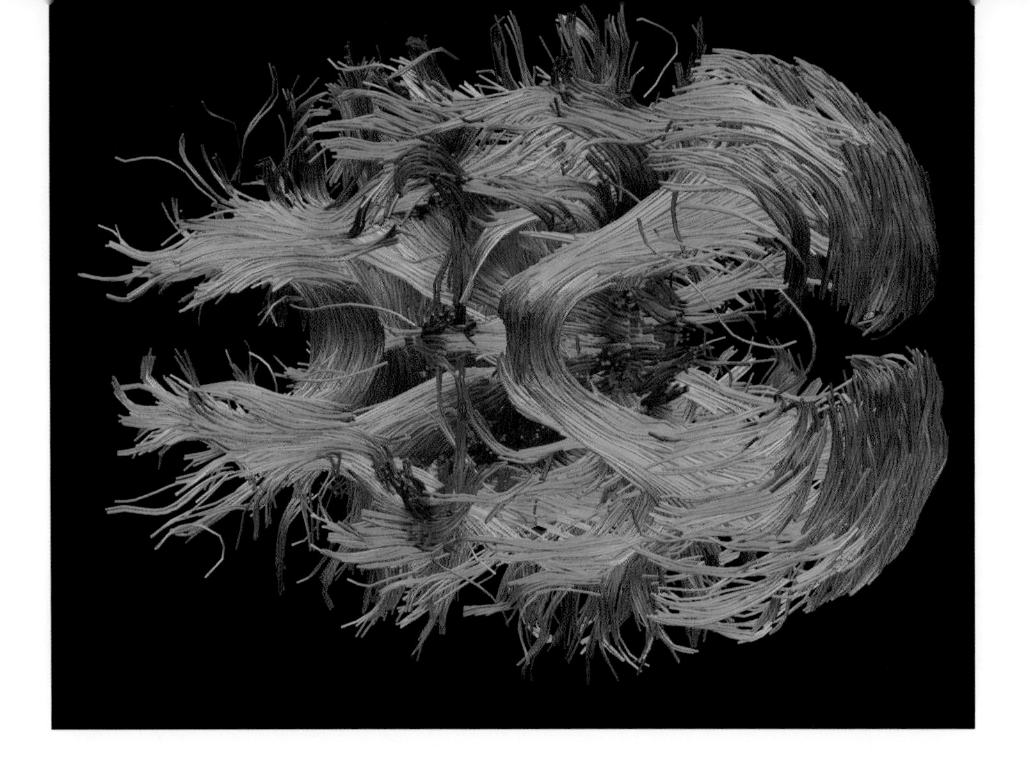

A genetic mutation is an alteration in the normal sequence of a gene that leads to a dysfunction.

A synapse is the area between two neurons that enables one to pass information to the next.

But it was only in the early 2000s with the development of molecular biology and sequencing techniques that we really began to understand the genetic mechanisms behind autism and to identify the mutations involved. In 2003, Thomas Bourgeron's team at the Institut Pasteur was the first to identify mutations in two genes on the X chromosome in two brothers, one with autism and the other with Asperger syndrome. These mutations affect genes that are directly involved in the functioning of synapses, or contact points between nerve cells. This was the first time that a link had been clearly identified between a genetic mutation, a disruption in synaptic function and autism spectrum disorder. Four years later, the same team identified another gene, this time actually involved in the structure of synapses. Based on research on a group of 227 children with ASD, the Institut Pasteur's scientists discovered various mutations affecting this gene. Their results confirmed the role of synaptic pathways in autism etiology and were subsequently backed up by US-based studies. But the story did not end there... In 2008, the same Institut Pasteur scientists discovered that some patients had a mutation in a gene involved in melatonin synthesis. This hormone plays a key role in regulating biological rhythms, especially sleep. Several children with autism have a melatonin deficiency, which may partly explain the sleeping difficulties experienced by approximately 60% of ASD patients. A genetically determined melatonin deficiency therefore appears to be another risk factor for autism.

Reconstruction of brain connectivity in white matter obtained using data from diffusion-weighted imaging.

Several factors involved

Since then, several other research teams across the world have identified further genetic mutations in ASD patients. More than 800 mutations have currently been identified, each time in just a handful of cases. These genes are involved in a large number of biological processes, although most of them play a role in the formation of the nervous system by modulating synapse formation (synaptogenesis) and gene regulation.

It is now generally accepted that genetic factors play a key role in vulnerability to autism. But the extreme variability of the disorders and the high number of mutations discovered mean that scientists have a difficult task ahead of them. While in some cases a mutation in a single gene can explain the majority of symptoms experienced by an ASD patient, in others the genetic situation is much more complex, sometimes involving more than a hundred genes that would not individually have had an impact, but which taken together increase the risk of being affected by some form of ASD. And the situation is complicated even further by the fact that in 10 to 20% of ASD patients, these genetic mutations are not inherited but occur *de novo* (spontaneously).

The promise of personalized medicine

These discoveries do nevertheless open up new avenues for treatment of children with ASD, especially in terms of medication. Close collaboration between geneticists, neurobiologists and psychiatrists has helped improve our understanding of how the brain works and shed light on the consequences of specific genetic mutations. Some teams are already testing tailored therapeutic solutions designed on the basis of a genetic mutation identified in a group of patients. Personalized medicine could therefore be a promising avenue for the treatment of patients with ASD.

Genetic research on autism in the Human Genetics and Cognitive Functions Unit directed by Thomas Bourgeron.

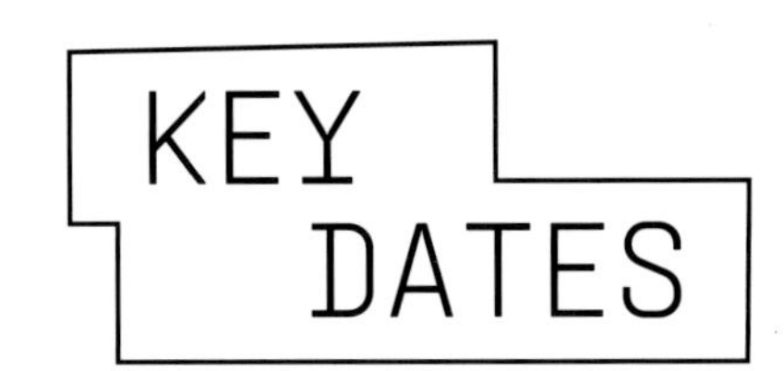

1911

1911: The term "autistic" is used in a psychiatry treatise for the first time. It is associated with schizophrenia.

1943

1943: American child psychiatrist Leo Kanner defines the clinical profile of early infantile autism.

1944

1944: Austrian child psychiatrist Hans Asperger uses the term "autism" to describe patients with difficulties interacting socially but normal language skills and cognitive function.

1950-1960

1950-1960: Based on his experience in concentration camps, American psychoanalyst Bruno Bettelheim puts forward the theory that the behavior of some "refrigerator mothers" is responsible for their children's autism.

1960-1970

1960-1970: Most experts believe that autism is a psychosis, an acquired mental disorder that should be treated using psychoanalysis.

1970-1980

1970-1980: American and French scientists use encephalography and advance the theory that autism is linked to early difficulties in sensory perception. A growing number of studies start to look at autism in twins or children from the same family. Experts begin to suspect that there may be a hereditary component to the condition.

2001

2001: A US report concludes that the best treatment approach for autistic children is individualized education programs. In France, psychotherapy is still the preferred treatment option.

2003

2003: For the first time, specific genetic mutations are identified in autistic children by Thomas Bourgeron's team at the Institut Pasteur, in cooperation with psychiatrists Prof. Marion Leboyer from France and Prof. Christopher Gillberg from Sweden. Other mutations are subsequently identified.

2005-2007

2005-2007: The first autism plan is introduced in France.

2008

2008: The first animal models with mutations associated with ASD are developed.

2008-2010

2008-2010: Second autism plan in France.

2013

2013: The fifth version of the Diagnostic and Statistical Manual of Mental Disorders (DSM-5) is published. It contains the diagnostic criteria for autism spectrum disorder.

2013-2017

2013-2017: Third autism plan in France (early detection and schooling for children with ASD).

2018-2022

2018-2022: Fourth autism plan under development.

Biology in the Age of Big Data

When biologists first suggested the idea of sequencing the entire human genome in 1985, it seemed like a monumental undertaking (see page 128). Experts believed that it would take 30 years to read all the genetic information contained by our 46 chromosomes—and with good reason: this information is stored in a very long molecule, DNA, which is composed of 3.2 billion "letters", or small molecules known as nucleotides that are represented by the letters A, T, G and C. At the time, methods for reading DNA were in their early infancy—the first results had been obtained just a few years earlier, in 1977—and when the genome of the Epstein-Barr virus (responsible for a form of herpes), which contains over 170,000 nucleotides, was sequenced in 1984 it was seen as a huge achievement. But despite the apparent obstacles, an international consortium of six countries, including France, embarked on the task in 1990. In 2001, they released a sequence of virtually all the human genome. The total cost of the project was in the region of $3 billion.

Genome: all the genetic material of a living organism. The genome is present in every individual cell of all living organisms. In humans, it comprises 23 pairs of chromosomes composed of a long coiled molecule known as DNA. If unraveled, the DNA in a human cell would measure nearly 2 meters.

Opposite: Capillaries from an Illumina cBot machine for the preparation of sequencing flow cells.

The “omics” era

Since then, things have got a lot faster. The 1000 Genomes Project, launched in 2008 with the aim of sequencing the genomes of a thousand people from all over the world, resulted in the sequencing of approximately 2,500 genomes in seven years. And with new high-throughput sequencing techniques, today we can sequence the genome of any person in just a few days for less than €1,000. We have entered the genomics era—or, more broadly, the “omics” era, in which techniques such as transcriptomics, proteomics, epigenomics and metabolomics can be used to obtain tens of thousands of data points in a single analysis of a given sample such as a cell, a cell population or an organ. The current sequencing output can be as high as 60 billion nucleotides per day, and biologists have begun analyzing larger samples, such as the microbial population in a drop of water, a handful of earth or the intestines of a cohort of patients.

Genomics: sequencing genomic data from a sample, whether a cell, an organism or the microbe population in a drop of seawater.

Transcriptomics: analysis of messenger RNA, the intermediary for protein production, which is produced in a cell in specific conditions (e.g. with and without drug treatment).

Proteomics: analysis of all proteins synthesized by a cell or a cell population.

Epigenomics: analysis of all changes to the genome that do not affect the DNA sequence.

Metabolomics: identification of the metabolites produced by a cell or a series of cells.

Mathematical modeling and advanced statistics and computing techniques have become vital in managing and analyzing the reams of data generated. Networks such as the European ELIXIR infrastructure have been set up with the aim of storing these resources in databases that can be accessed by scientists, as well as developing tools and training methods so that they can be used effectively. The French branch of ELIXIR, the French Institute of Bioinformatics (IFB), is an umbrella organization for around twenty platforms across the country, under the leadership of public research organizations (the CNRS, Inserm, INRA, Inria and the CEA), universities, the Institut Curie and the Institut Pasteur.

100,000 HUMAN GENOMES... AND 1,000 BRETON GENOMES

When the scientists working on the 1000 Genomes Project finally obtained the results of their initial data analysis in the early 2010s, they realized that the genomes they had chosen were too distant from each other to provide information about some genetic diseases. Recent projects have therefore taken a more targeted approach. The Institut Pasteur launched a project entitled “1000 French Genomes” (mostly from Brittany), taking a cohort of a thousand people in the city of Rennes with the aim of studying variations between individuals and the potential links with infectious diseases or response to treatment. The first results are expected in 2018. In Asia, a large-scale project was launched in 2016 to analyze the genomes of 100,000 people from the entire continent to help shed light on the onset of rare congenital and chronic diseases.

Opposite: Photo taken in the Transcriptome and EpiGenome Platform in 2013.

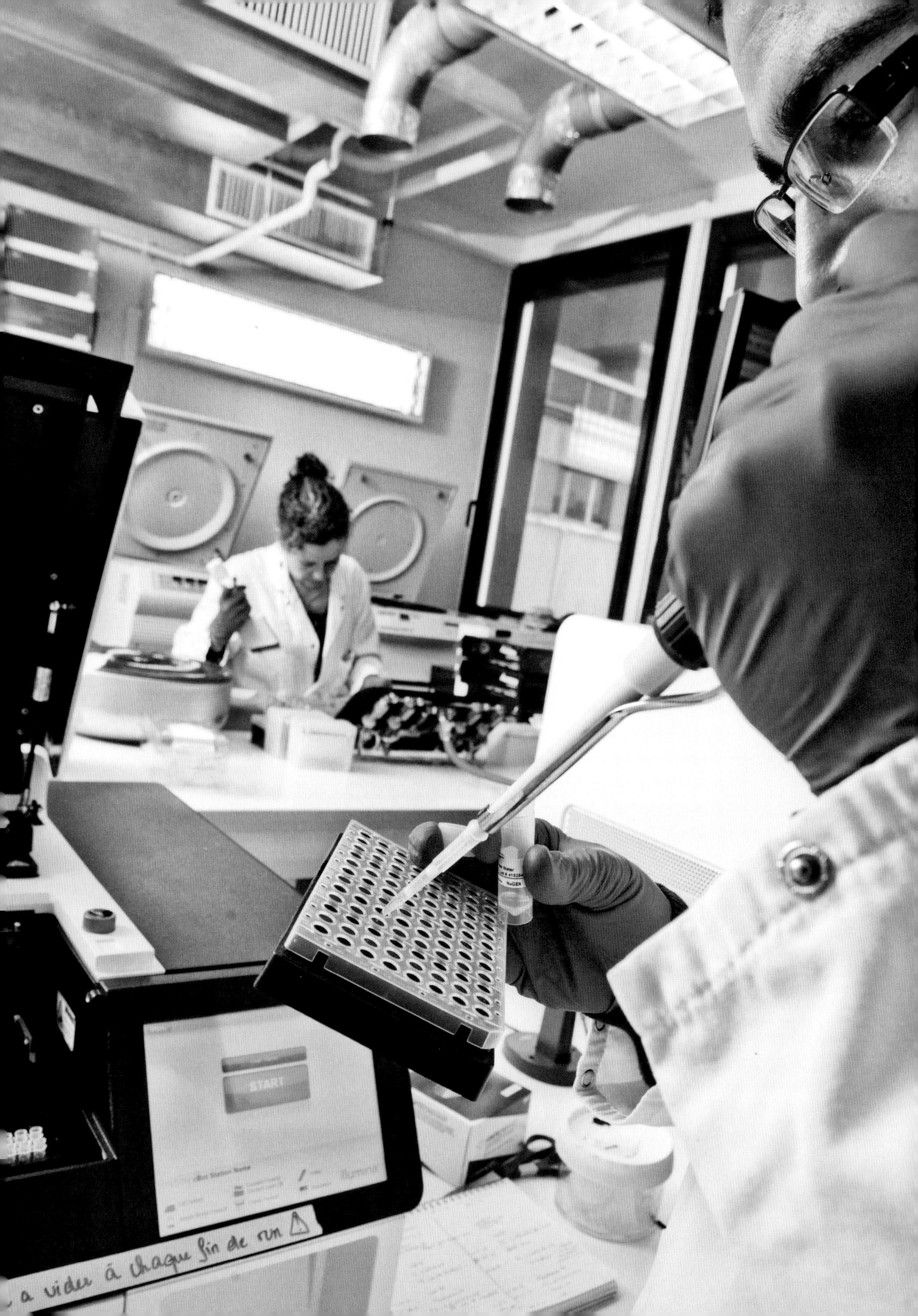
START
a vider à chaque fin de run

One of these platforms is based at the Institut Pasteur in Paris, in the Center of Bioinformatics, Biostatistics and Integrative Biology (C3BI) directed by Olivier Gascuel. With around 40 engineers, 3,500 computing cores and a storage capacity of 2,200 terabytes (1 terabyte equals 10^{12} bytes), it is one of the best-equipped platforms in the IFB. But the C3BI's ambitions go beyond the national framework: it is aiming to become a leading international center for biology data processing, analysis and modeling. While considerable progress has been made worldwide in recent years in terms of data acquisition, with a variety of applications in areas such as health care and biodiversity research, progress in data analysis has tended to lag behind. Scientists have also realized that the cost of data analysis is far higher than for data acquisition. This has led the Institut Pasteur to focus its efforts on developing methods for analyzing and making use of large-scale data.

"Since 2016, the C3BI has been affiliated to three CNRS institutes in the areas of biology, IT and environmental studies, which facilities cross-disciplinary dialog between specialists in the various fields."

Olivier Gascuel, Director of the Center of Bioinformatics, Biostatistics and Integrative Biology (C3BI) at the Institut Pasteur

THE C3BI, A CENTER DEDICATED TO DIGITAL BIOLOGY

The Center of Bioinformatics, Biostatistics and Integrative Biology (or C3BI) is one of the Institut Pasteur's four transversal research centers. These centers were set up in 2014 and 2015 to facilitate contacts between scientists at the Institut Pasteur and with other organizations, and to encourage the transfer of knowledge and skills. The C3BI facilitates cooperation and dialog in bioinformatics at the Institut Pasteur, and it also offers a high-level service to scientists via its Bioinformatics and Biostatistics HUB. These services are also available for the Institut Pasteur International Network (IPIN) and at European level through the ELIXIR network. In addition to these services, the C3BI develops tools and methods in computing and statistics and provides training in bioinformatics for IPIN researchers. Scientists can come to the C3BI for guidance on how to analyze data generated in their laboratories, or in collaboration with one of the Citech platforms (see opposite) or a sequencing center.

Element of the bioinformatic cluster computing infrastructure available for use by Institut Pasteur scientists to process and analyze their data.

From bioinformatics to systems biology

The C3BI, which currently has 14 teams and 132 members, was launched in March 2015 as a multidisciplinary center.

The three teams in the bioinformatics platform offer support for the Institut Pasteur's scientists by providing tools and training to help them manage and analyze their data. In just two years, 190 project proposals have been submitted and more than 60 have already been completed on topics including sequencing the genome of pathogens, comparing yeast genomes, analyzing the transcriptome of the parasite responsible for malaria, identifying new disease-causing microbes, analyzing the polymorphism of the seasonal influenza virus, and investigating the gut microbiota (the microbes living in the intestinal tract) in children.

The other units, some of which existed before the C3BI was set up and some which were created at the same time, are developing new research areas in bioinformatics, biostatistics and modeling. With the arrival of "omics" approaches in the 2000s, a new discipline, systems biology, began to emerge in several countries across the world, with the aim of shedding light on the workings of complex biological systems such as a cell, a group of cells or a network of proteins. The idea is to use all the data accumulated about a given system, for example all the interactions between its proteins, and to predict

THREE OTHER CENTERS WORKING ALONGSIDE THE C3BI

The Citech (Center for Innovation and Technological Research) houses 13 platforms, the Central Animal Facility and a new technology laboratory. It provides scientists at the Institut Pasteur and in the International Network with access to cutting-edge equipment and methods. They approach the Citech when they need genomic data or data from other "omics" techniques.

The Center for Translational Science (or CRT) was set up to facilitate biomedical research projects between members of the Institut Pasteur International Network and clinicians. The CRT develops partnerships with hospitals so that scientists have access to patient cohorts. It also assists scientists looking to develop clinical trials, for example by helping them secure funding. Lastly, it offers training and provides two technical platforms for medical research, the Center for Human Immunology and the Cytometry Platform.

The Center for Global Health (CGH) provides support and visibility for Institut Pasteur projects that aim to tackle global health threats, by drawing on the Institut Pasteur's research and education activities. It has set up an Outbreak Investigation Task Force (see page 92). It also promotes the wide-ranging expertise in the Institut Pasteur International Network on global health issues (antibiotic resistance, child malnutrition and the fight against dengue), working closely with the Institut Pasteur's Department of International Affairs, which coordinates the network. Finally, the CGH offers training for scientists and clinicians, for example in Africa with the Pan-African Coalition for Training in Research and Public Health (PACT).

its behavior using IT and mathematical modeling. In 2012, Stanford University's Jonathan Karr and his colleagues published a model of the workings of the *Mycoplasma genitalium* bacterium, incorporating all its constituent parts and their interactions. This model enabled the scientists to predict how the bacterium would behave (in other words the proteins it would produce and their activity) and to validate their predictions against independent experimental data. There are a host of potential applications for these techniques, such as testing out hypotheses on a modeled system, predicting how a tumor will respond to a given treatment depending on its characteristics, and highlighting properties that are not revealed during experiments.

In March 2017, the C3BI and the French National Institute for Computer Science and Applied Mathematics (Inria) set up a dedicated structure named InBio to explore the potential of this strategy. But much of the C3BI's research activity takes an approach that is more closely aligned with the Institut Pasteur's research areas: the center's main focus is on all aspects of evolution, including the spread of disease, human evolution, and the evolution of microbes and their resistance.

In this research field, biologists approach the C3BI teams with questions on a wide range of topics including biodiversity, viruses and epidemics. And very often there are IT and mathematical tools that can be used to find the answers. For example, mathematical results that have served as a basis for modeling future biodiversity scenarios are currently being used to investigate the spread of resistance in HIV strains.

Understanding the emergence of disease

A flagship project in this area, INCEPTION—which received a ten-year grant of €12 million under the French Government's Investing in the Future Program in 2016—was set up to carry out research into the emergence of infectious diseases (such as Ebola, Zika, AIDS, etc.) and other conditions whose causes are often poorly understood, such as cancer and autism. Étienne Simon-Lorière from the Functional Genetics of Infectious Diseases Unit, which helped characterize the Ebola virus in 2016 (see page 98), has recently joined the INCEPTION project to carry out research on this virus and the mutations it has acquired over the course of its evolution.

The Mathematical Modeling of Infectious Diseases Unit, directed by Simon Cauchemez, focuses on more epidemiological aspects. Scientists in this unit analyze human behavior and the environmental and social contexts that may favor the spread of a given virus. Using epidemiological data on

Stage in the high-throughput DNA sequencing procedure. Here, an automatic thermal cycler is used to generate clusters of sequences on a slide known as a flow cell.

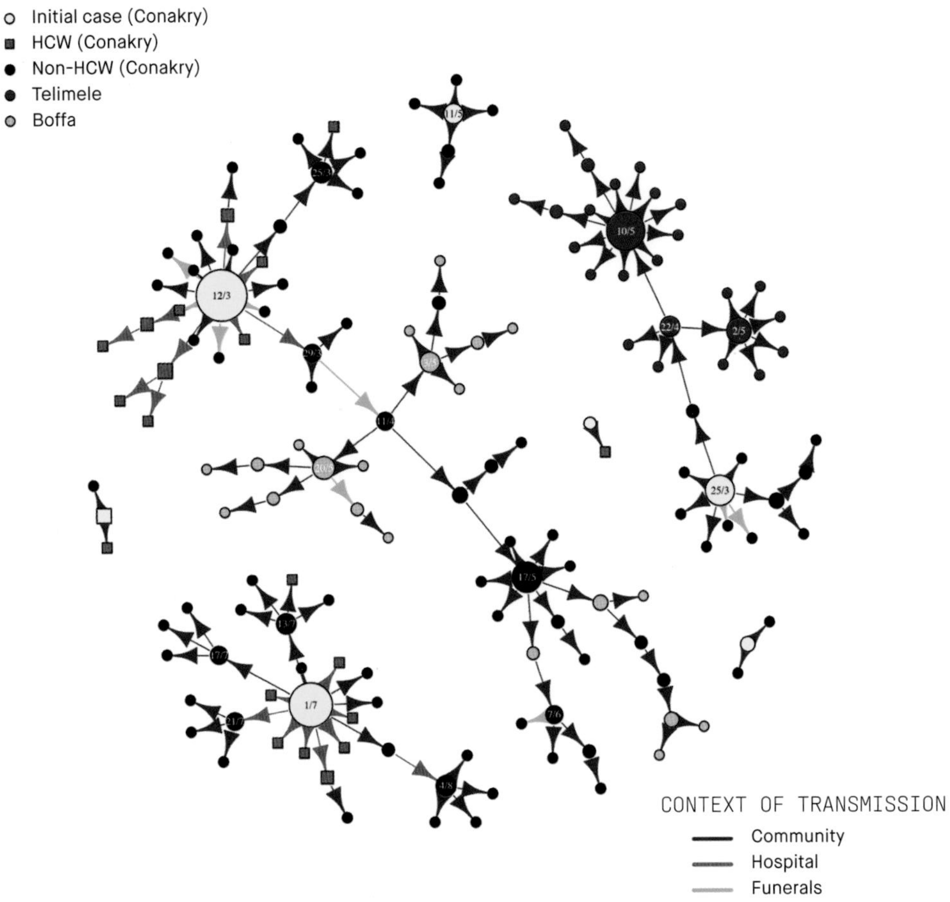

Initial case (Conakry)
HCW (Conakry)
Non-HCW (Conakry)
Telimele
Boffa

CONTEXT OF TRANSMISSION
Community
Hospital
Funerals

Ebola, influenza, dengue, chikungunya, Zika and yellow fever collected in a number of countries, the team is developing methods to improve understanding of the spread of infectious diseases in humans so that we can better anticipate and prepare for health emergencies. The aim is to identify factors influencing the spread of pathogens so as to improve the health strategies adopted to monitor and control outbreaks.

Transmission trees for the Ebola virus. Dates of symptom onset are indicated in circles for cases that infected more than three people. The size of the circles is proportional to the number of people infected by the case.

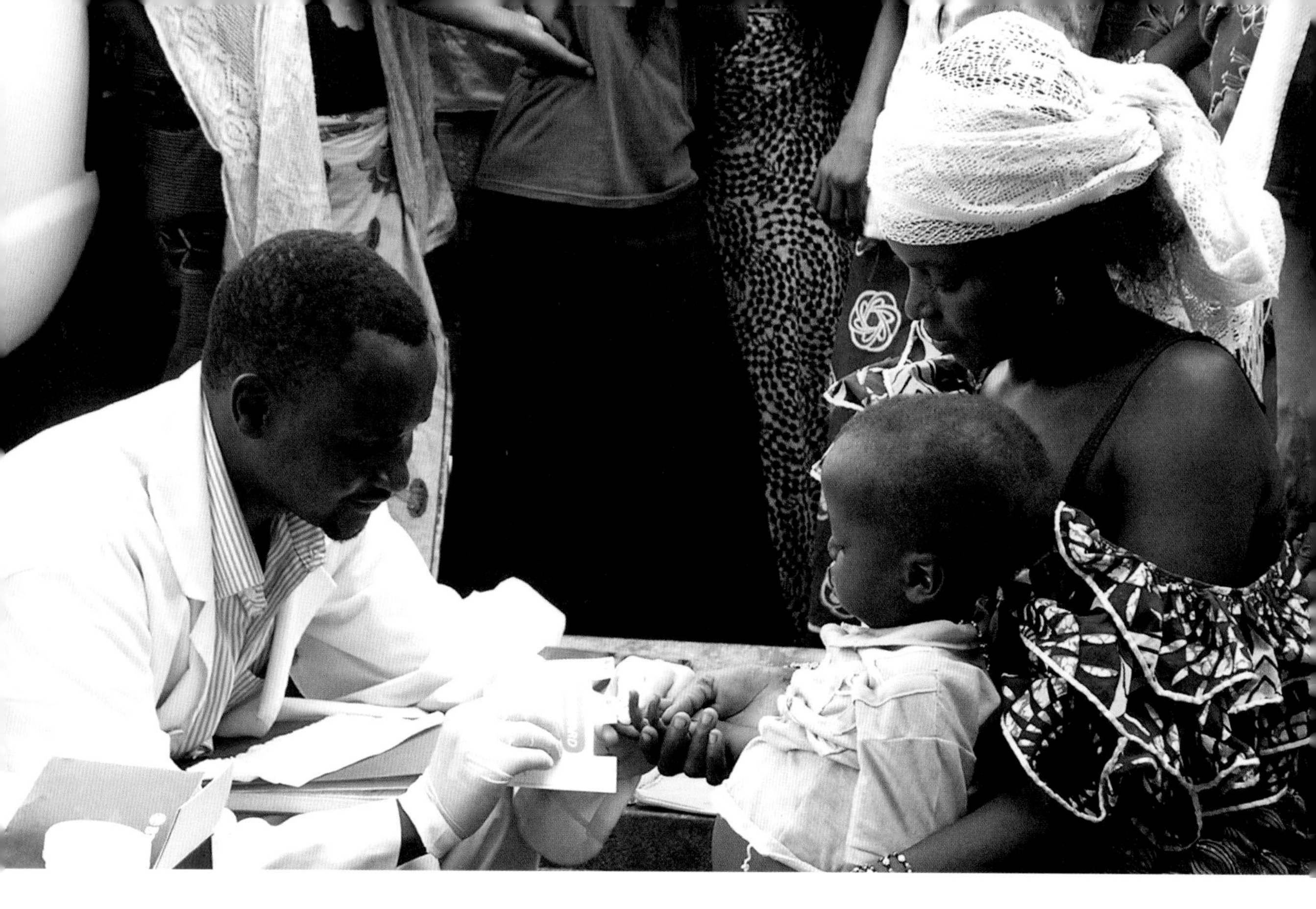

Mathematical modeling can provide a wide range of information about diseases. For example, by analyzing data from French Polynesia, Simon Cauchemez and his colleagues were able to provide one of the first estimates of the risk of microcephaly associated with Zika virus infection in pregnant women (see page 104). In a joint project with Santé publique France, they also developed mathematical models to predict the development of the Zika epidemic in Martinique and evaluate the hospital resources that would be needed for effective treatment of patients with more severe complications. And when a yellow fever outbreak hit Angola and the Democratic Republic of Congo, the team worked with colleagues from the University of Oxford to develop mathematical models to better anticipate the spatial propagation of the virus. This work was based on an analysis of several datasets providing information on factors such as population density and mobility, as well as climate and socio-economic conditions that would be more or less conducive to the virus spreading via its main vector, the *Aedes aegypti* mosquito.

Monitoring the epidemiology of malaria in Niger in 2004.

"In situations where there are limited vaccine supplies, this type of study can help decision-makers to optimize vaccination campaigns."

Simon Cauchemez, Head of the Mathemetical Modeling of Infectious Diseases Unit at the Institut Pasteur

Microbes and their hosts

Other approaches adopted at the C3BI involve studying infectious agents themselves and looking at their biology from an evolutionary perspective. The aim is to understand how microbes evolve when they come into contact with their congeners, their hosts and their environment, and especially how they develop resistance to treatment (see page 172). But research into infectious diseases would be lacking if it did not also consider the human host. We are not all susceptible to diseases to the same extent: the immune response to infection varies from one person to another depending on their genetic makeup, environment and background. At the C3BI, the Human Evolutionary Genetics Unit directed by Lluis Quintana-Murci is therefore exploring the evolution of the immune system in humans by studying population genetics.

Gene: an element of the genome that carries a precise inherited characteristic in the form of a coded instruction. The cell reads this code and follows the instruction, which leads to the production of a specific protein. The human genome has approximately 35,000 genes.

The scientists in this unit look at the distribution of different forms of some genes in a range of human populations and examine how this distribution develops over time. Some forms make us more vulnerable to pathogens, while others protect us from infection, giving individuals carrying them an advantage. The aim is to predict how a given person will respond to infection or vaccination so that treatments can be adapted to be as effective as possible. Lluis Quintana-Murci and his colleagues have identified innate immunity genes whose sequence does not vary from one individual to another, suggesting that they are essential for the body's defense—if no mutation of these genes exists in nature, it is probably because such mutations would lead to severe or even fatal diseases.

In 2016, the team showed that we owe part of our immune system to Neanderthals. We have known for some years that early modern humans interbred with Neanderthals around 40,000 to 50,000 years ago, and that we share approximately 3% of our DNA with Neanderthals. In collaboration with teams from the French National Genotyping Center (the largest French

Artwork created by Jérôme Bon for the Institut Pasteur illustrating the genetic legacy passed on by Neanderthals to Europeans and its influence on their ability to protect themselves from viruses.

sequencing platform, based at the Évry Genopole), Ghent University and the Leipzig-based Max Planck Institute for Evolutionary Anthropology, Lluis Quintana-Murci's team decoded the immune response of 200 individuals from Africa and Europe. Their analysis revealed major differences in the way these individuals respond to infections, and indicated that these variations are largely attributable to genetics. More striking still, some innate immunity genes contain more Neanderthal DNA than the rest of the human genome. This shows us that these Neanderthal sequences conferred an advantage on the humans carrying them and were gradually selected during evolution.

These examples of research give us just a glimpse of the immense possibilities afforded by bioinformatics and the "omics" era. And this exciting new wave shows no sign of stopping any time soon.

> "By analyzing the genomes of 80 bacterial species, we demonstrated that more than 60% of the genes they receive from other bacteria are integrated into just 1% of the genome regions."
>
> Eduardo Rocha, Head of the Microbial Evolutionary Genomics Unit

HOW BACTERIAL RESISTANCE EMERGES

Most of the time, bacteria become resistant to a harmful environment (such as an antibiotic treatment—see page 60) by acquiring genes from other bacteria via the transfer of genetic sequences, rather than because of a random mutation that has occurred in their genome. The Microbial Evolutionary Genomics Unit, directed by Eduardo Rocha and affiliated to the C3BI, is working to understand how these mobile genetic elements integrate into the bacterial genome, thereby adding new genes. One focus of their research is how the genomic organization of bacteria can handle such an influx of information. By studying the genome evolution of a bacterial species, Eduardo Rocha and his colleagues observed that this is a highly organized process. Harmful mutations soon eliminate the bacteria carrying them, meaning that the bacterial chromosome is under high selective pressure: only genes that confer a definite advantage on bacteria will survive. But mobile genetic elements are constantly bombarding bacteria with a considerable volume of new information. How do they manage this influx? Eduardo Rocha's team has recently revealed part of the answer: only some regions of the genome receive information that is likely to make the bacteria resistant to a new environment.
The team had already observed this phenomenon in the *Escherichia coli* bacterium a few years earlier, but this latest research proves that it is widespread. The biologists are currently investigating in more detail the tools used by the bacterium to regulate the arrival of these mobile elements. They are also trying to determine whether interference between the various known transfer mechanisms, each of which has its own dynamic, speeds up or slows down the acquisition of resistance genes. Finally, in clinical trials with a team from Bichat Hospital, they are examining how resistance to an ingested antibiotic spreads within the microbial communities in the intestine. To be continued...

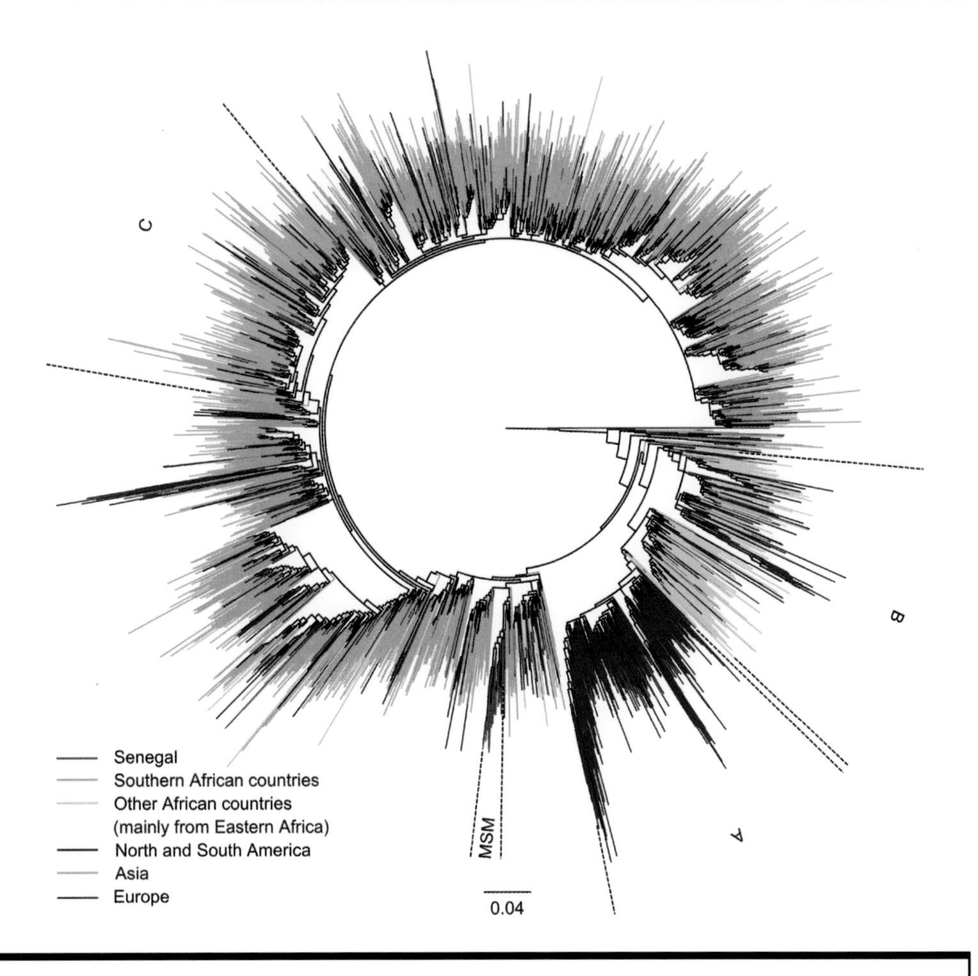

THE TENTH HIV GENE

When the Evolutionary Bioinformatics Unit directed by Olivier Gascuel confirmed the existence of a tenth HIV gene, it was a clear illustration of how IT can help advance biology. Since 1988, the existence of this gene had been shrouded in controversy. The HIV genome is a molecule composed of a single strand of RNA (RNA is an intermediary in protein production). When HIV infects a cell, the cell "retrotranscribes" the viral RNA as DNA and incorporates this DNA into its own genome. In 1988, scientists discovered a sequence that had all the characteristics of a gene in the DNA produced from viral RNA. But many biologists remained skeptical, even after it was demonstrated more recently that this gene is translated into a protein *in vivo* and that this protein triggers a response from the human immune system. They claimed that there were plausible explanations for all of these phenomena.
However, in 2016, Olivier Gascuel and his colleagues provided additional proof when they analyzed the evolution of the virus. By analyzing all the sequences obtained across the world for this part of the HIV genome, they showed that viral RNA corresponding to the tenth gene only appeared in M-group strains, which account for virtually all the cases in the global pandemic (M meaning Major). Moreover, the team demonstrated that this gene has a particularly well-conserved sequence, indicating that it gives the virus a major transmission advantage. The gene could therefore explain why the M group caused a pandemic, whereas the other three groups (N, O and P), which are lacking the gene, did not have the same effect.

(For further information about AIDS, see page 78.)

This phylogenetic tree, produced using IT methods, shows how HIV-1 subtype C was introduced into the world. Sometimes infection with this subtype can be traced back to (virtually) a single introduction—in Asia, America and Senegal among MSM (men who have sex with men)—, while in Europe, for example, there were multiple introductions.

Beyond Research

Medicine, Public Health and Surveillance

EXPERTISE SUPPORTING GLOBAL HEALTH

16,000
pathogen sequences (bacteria, viruses or fungi) completed in 2016

14
National Reference Centers (CNRs), in January 2017

6
WHO Collaborating Centers (WHOCCs), in January 2017

The Institut Pasteur's prominent role in health surveillance

In the late 19th century in France, tens of thousands of children died each year from diphtheria. As soon as the Institut Pasteur was set up in 1888, two of Louis Pasteur's close collaborators Émile Roux and Alexandre Yersin began investigating how to treat these children, many of whom were admitted to the Children's Hospital located between Rue de Sèvres and Rue de Vaugirard just a stone's throw from the Institut Pasteur. They discovered diphtheria toxin and shortly afterwards their German counterparts developed an antitoxin. This was a clear example of how scientific expertise could help tackle a public health problem, and it confirmed the Institut Pasteur's humanist and universal vocation to improve people's health through the development of resources to conduct disease surveillance and monitoring activities.

1 The Institut Pasteur's articles of association provided for "the creation and management of laboratories for referral, testing and monitoring, and collections of microbial strains", right from the start.

National Reference Centers (CNRs)

Since the 1950s, the French government has coordinated efforts to monitor public health, especially with regard to infectious diseases, through National Reference Centers, or CNRs (their official name since 1972). The country's CNRs currently operate under the aegis of Santé Publique France (formed from the merger between the French Institute for Public Health Surveillance, or InVS, and two other public health organizations). Of the 44 CNRs operating in France in 2017, 14 are hosted by the Institut Pasteur. There are also four CNR-associated laboratories based at the Institut Pasteur in French Guiana.

The centers serve as infectious disease observatories—their mission is to identify and confirm the culprit in suspected cases of infection. They rely on networks of doctors and laboratories to notify them directly.

The work of each CNR focuses on one specific infectious agent, for example *Listeria*, the rabies virus or respiratory viruses (like influenza). In practice, surveillance activities are carried out by heads of research units. CNRs use a variety of molecular and genetic analysis techniques to identify precisely which pathogenic strains are circulating (bacteria, viruses or other pathogens), and they also characterize their resistance to treatment. They report any unusual events to the health authorities. This may include the emergence of a new strain of bacterium, virus or other pathogen, the appearance of unusual or particularly virulent clinical signs of a disease, a cluster of cases that might develop into a more widespread epidemic, or the emergence of resistance to antibiotics or antiviral drugs. CNRs also participate in the training of decision makers and other public health workers.

National Reference Centers (CNRs), which serve as effective observatories for communicable diseases and expert microbiology laboratories, depend on the Institut Pasteur's scientific expertise. Here, a view of the Institut Pasteur campus, showing the building housing the François Jacob research center, which opened on November 14, 2012 (on the left).

FOUR CNRs HOSTED BY THE INSTITUT PASTEUR

MONITORING SEASONAL OUTBREAKS AND PANDEMIC RISKS

THE NATIONAL REFERENCE CENTER FOR RESPIRATORY VIRUSES INCLUDING INFLUENZA

Every winter, this Institut Pasteur laboratory keeps its finger on the pulse of the country's seasonal influenza outbreak. When did it start? Which viral strains are circulating among the public? Does the vaccine match these strains? The CNR analyzes around 50 samples every day from family physicians and hospitals nationwide. The aim is to characterize influenza strains as quickly as possible, as well as other viruses that cause influenza-like illness and acute respiratory infections, such as respiratory syncytial virus (RSV), the most common cause of infant bronchiolitis. Through their constant vigilance and the information they collect on strains in circulation, every year CNR experts, along with other European experts, are called on to advise on the best possible composition of the vaccine to be manufactured for the following season. Also, in suspected cases of infection caused by an animal strain, avian influenza for example, the CNR works with teams specializing in epizootic monitoring to characterize the strain responsible and determine the likelihood of a pandemic.

Epizootic: a disease outbreak in a non-human species.

CONFIRMING SUSPECTED CASES

THE NATIONAL REFERENCE CENTER FOR MENINGOCOCCI AND *HAEMOPHILUS INFLUENZAE*

Complaints of a stiff neck and vomiting among schoolchildren inevitably lead to panic. Meningitis strikes such fear because it is highly infectious and can be fatal, especially in very young children—which means that it is vital to act quickly. Every year in France, there are 500 cases and 30 deaths in people under the age of 20. Meningitis can be caused by one of many meningococcal bacteria which infect the membrane around the brain or cause septicemia (a widespread infection of the blood and organs). These diseases are notifiable in France. Whenever there is a suspected case, the Institut Pasteur laboratory is approached to provide biological confirmation of the diagnosis, which then enables the health authorities to take emergency measures to prevent the disease from spreading. The CNR also has the task of monitoring the emergence of new bacterial strains. Additionally, since the development of vaccines for some meningococcal strains, the laboratory has also been involved in analyzing whether they are effective against the strains circulating in France, and developing the most appropriate vaccine policy in cooperation with the health authorities.

Notifiable: these diseases must be reported to health authorities by the health care professionals that diagnose them.

IMPROVING UNDERSTANDING AND TREATMENT OF LISTERIOSIS

THE NATIONAL REFERENCE CENTER FOR *LISTERIA*

Pregnant women are advised to avoid raw milk cheese, cured meats and even salads made with raw vegetables because they might be contaminated with *Listeria monocytogenes*. This bacterium is responsible for a dangerous infection, listeriosis, that can cause septicemia or a central nervous system infection, and if carried by pregnant women, miscarriage, preterm labor, or a severe neonatal infection.
Since 1982, procedures have been introduced in France to identify and closely monitor cases of listeriosis. While the disease remains rare, it is fatal in 20 to 30% of cases outside of pregnancy. Since 1993, the CNR has been responsible for epidemiological surveillance through the characterization of bacterial strains isolated in patients. Recently, scientists at the Institut Pasteur published an extensive review of the latest knowledge about the infection and clarified the clinical signs and prognostic factors, especially in pregnant women and elderly or vulnerable populations. They also revealed the potential of an antibiotic treatment that had not previously been included in the official medical guidelines. This research will help improve treatment for listeriosis patients.

RAPIDLY DETECTING CLUSTERS

THE NATIONAL REFERENCE CENTER FOR *ESCHERICHIA COLI, SHIGELLA* AND *SALMONELLA*

These three families of bacteria are responsible for a variety of diseases that mainly affect the digestive system—including dysentery, infectious diarrhea, hemolytic-uremic syndrome and foodborne infections—as well as typhoid and paratyphoid fever, and salmonellosis. The surveillance and alert system developed by the Institut Pasteur laboratory enables early detection of any unusual increase in the number of cases locally or across the country. The aim is to identify the emergence of an epidemic as quickly as possible, or to pinpoint the origin of a foodborne infection, for instance. The experts conduct a thorough investigation in which they precisely type the bacterial strains in several human biological samples and food samples to determine whether clusters of an infection, such as salmonellosis, have been caused by a contaminated food product. Depending on the bacteria, the Institut Pasteur teams also carefully monitor the emergence of any antibiotic resistance, and closely examine the mechanisms involved.

Top left: An influenza virus sample analyzed at the Influenza National Reference Center in winter 2012-2013.

Top right: A photo taken in the Molecular Genetics of RNA Viruses Unit, which houses the National Reference Center for Influenza Viruses.

World Health Organization Collaborating Centers (WHOCCs)

Beyond nationwide surveillance, changing lifestyles (increasing numbers of people traveling and the growing popularity of "exotic" destinations, for example) mean that diseases now need to be monitored on a global scale. The WHO management team selects laboratories around the world to become WHO Collaborating Centers (WHOCCs), creating an international network of laboratories tasked with testing and monitoring infectious agents to support WHO research and prevention programs. Six WHOCCs are currently based at the Institut Pasteur. Their tasks include the surveillance and detection of arboviruses and viral hemorrhagic fevers (including Ebola and the Marburg virus), rabies and resistance to antimalarial drugs.

Arbovirus: a type of virus that is transmitted by blood-feeding arthropods (vectors), e.g. mosquitoes and sand flies.

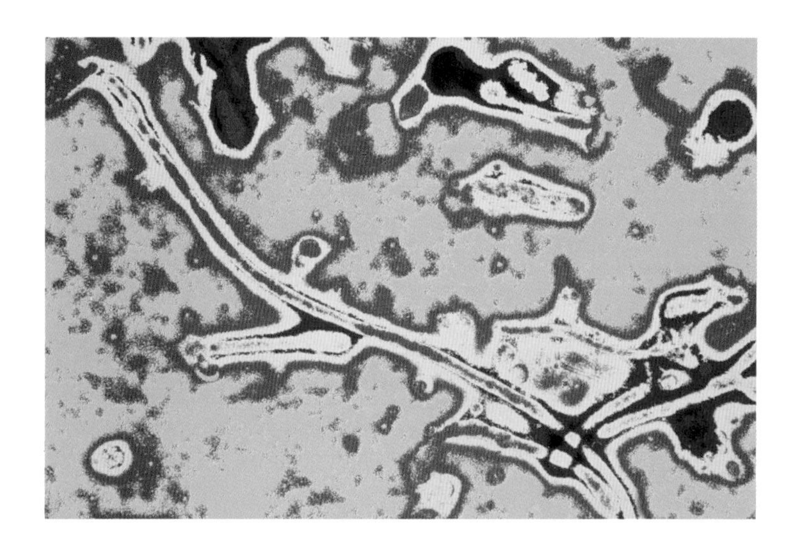

EBOLA: HOW THE VIRUS WAS IDENTIFIED

In December 2013, the first cases of hemorrhagic fever appeared in Guinea. A few weeks later, the WHO Collaborating Center (WHOCC) for Viral Hemorrhagic Fevers and Arboviruses, led by Noël Tordo at the Institut Pasteur, was notified by Médecins sans frontières. The WHOCC then notified Sylvain Baize, Director of the National Reference Center (CNR) for Viral Hemorrhagic Fevers at the Institut Pasteur, in Lyon. On March 21, 2014, the CNR confirmed that the causative agent was a filovirus (Ebola virus family). The next day, the Laboratory for Urgent Response to Biological Threats (known as CIBU, and managed by Jean-Claude Manuguerra) in Paris identified the pathogen as Zaire ebolavirus (ZEBOV).

The Ebola virus (family *Filoviridae*). Ebola is the longest-known filamentous virus, causing high fever and internal bleeding that can often prove fatal for humans and monkeys. Colorized image.

The Laboratory for Urgent Response to Biological Threats (CIBU) and the Institut Pasteur Biological Resource Center (CRBIP)

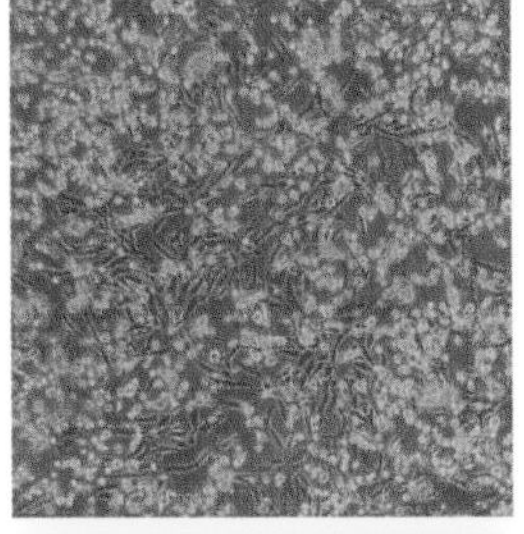

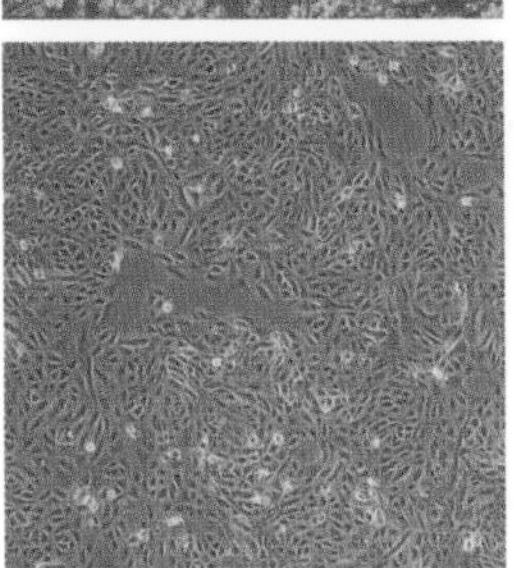

If an infant dies suddenly from a mysterious infection, there are reports of a cluster of outbreaks of food poisoning of unknown origin or a traveler returns ill from a trip to an exotic destination and is thought to be contagious, the Institut Pasteur's CIBU team leaps into action to respond to these emergencies, 24/7. It receives samples which it uses to rapidly detect and identify whether an infectious agent (a virus, bacterium or parasite) or a biological toxin is responsible. It also takes action when the health authorities suspect a public health threat that has been targeted in a specific public health plan, for example the bacteria responsible for anthrax or plague or the viruses that cause Ebola, influenza pandemics or SARS.

The Institut Pasteur also has its own "infection memory bank", a valuable tool that is vital for both scientists and public health stakeholders—the Institut Pasteur Biological Resource Center (CRBIP). This bank collects and preserves biological material including bacterial strains, cell cultures, genetic material, human biological samples and all related information. These samples, sometimes collected over several decades, serve to demonstrate changes that have occurred over time and to adapt strategies accordingly (effectiveness of antibiotics, changes to vaccine composition, etc.).

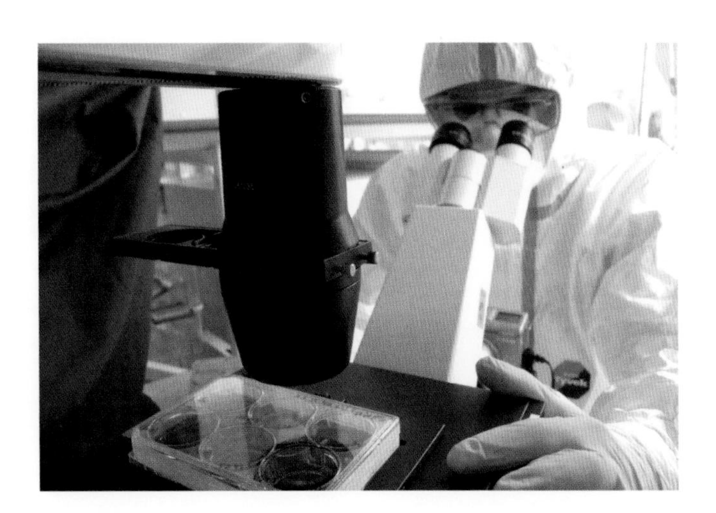

Top: The cytopathic effect of MERS-CoV. The top image uses phase-contrast photon microscopy to show the cytopathic effect of MERS-CoV after four days of culture. The bottom image, by contrast, is the negative control cultured in the same conditions.

Above: Inverted microscopy observation (BSL-3) in the Laboratory for Urgent Response to Biological Threats (CIBU).

A Medical Center at the Institut Pasteur

The Institut Pasteur Medical Center, open since 2000, is where the Institut Pasteur comes into direct contact with patients. If you have been bitten by a dog or suffer from allergies, need to be vaccinated for a trip overseas, want specific travel advice, or are suffering from an infectious tropical disease after spending time abroad, you can visit the Medical Center, which has been specializing in these areas ever since it took over from its forerunner, the Pasteur Hospital, founded back in 1900.

In 2015, more than 45,000 people visited the Medical Center's International Vaccination Center, which administered over 70,000 vaccines. As well as vaccinations, preventive advice for travelers and the information available at www.pasteur.fr, in 2015 the Medical Center also provided more than 12,000 consultations for infectious, tropical and skin diseases and more than 2,000 consultations following exposure to rabies. Some diseases, such as HIV, are monitored in cooperation with the Infectious and Tropical Diseases Department at Necker Children's Hospital.

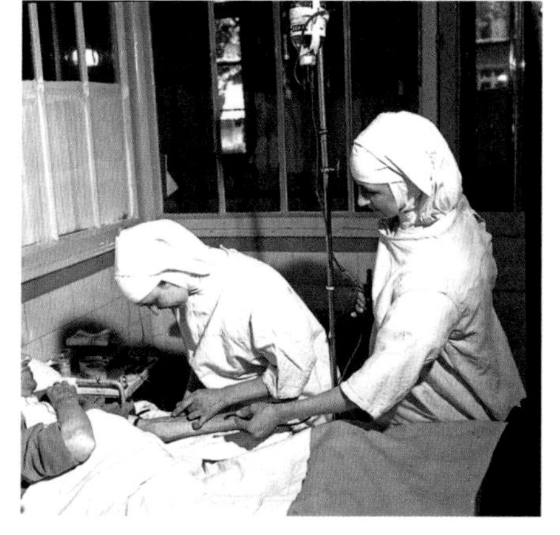

Patient isolation at the Pasteur Hospital, around 1950. Sister Reine (left) and Sister Damien (right).

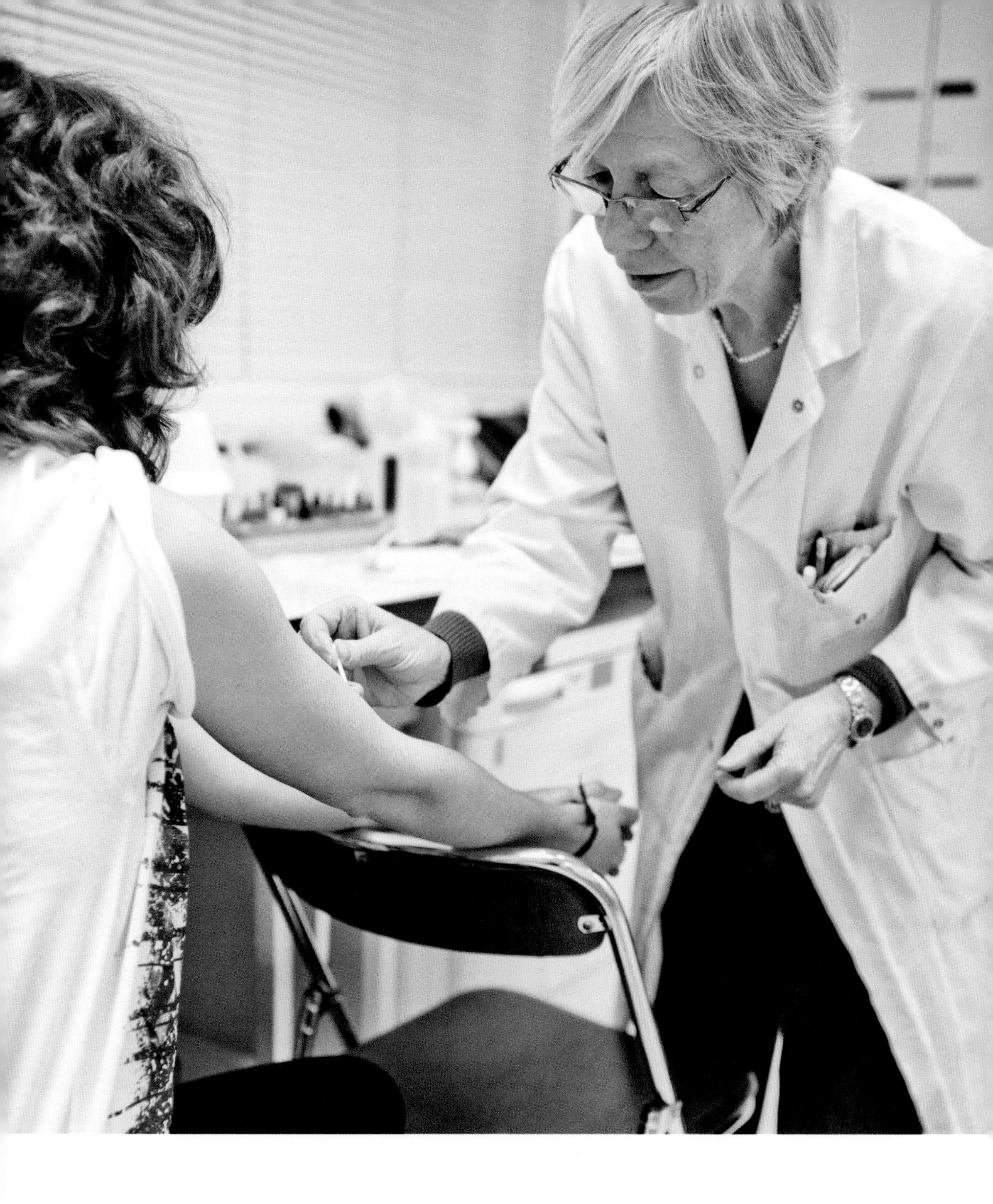

An allergy consultation at the Institut Pasteur Medical Center in 2013. Skin tests can help pinpoint the cause(s) of allergies.

The Medical Center is also involved in clinical research, monitoring cohorts of patients with HIV and hidradenitis suppurativa. In its severe form, hidradenitis suppurativa can be considered an orphan disease, and has become the focus of pathophysiological research undertaken alongside therapeutic treatment at the center. In 2015, this research revealed that for each stage of hidradenitis suppurativa severity, different bacterial flora can be found in the skin lesions it causes.

The Medical Center also conducts research on vaccine administration procedures. In 2016, it examined the interaction between two vaccines that are sometimes administered together: the yellow fever vaccine and the measles, mumps and rubella (MMR) vaccine. In a trial of 131 children aged 6 to 24 months, the yellow fever vaccine only failed to offer protection in children who had received both vaccines on the same day. Similarly, MMR vaccination failure only occurred in a few children who had been given the vaccines less than 28 days apart. These initial findings need to be confirmed by further research, but they could suggest that, contrary to current guidelines, the efficacy of each vaccine would be optimized if the two were not administered on the same day, and that they should ideally be given more than 28 days apart.

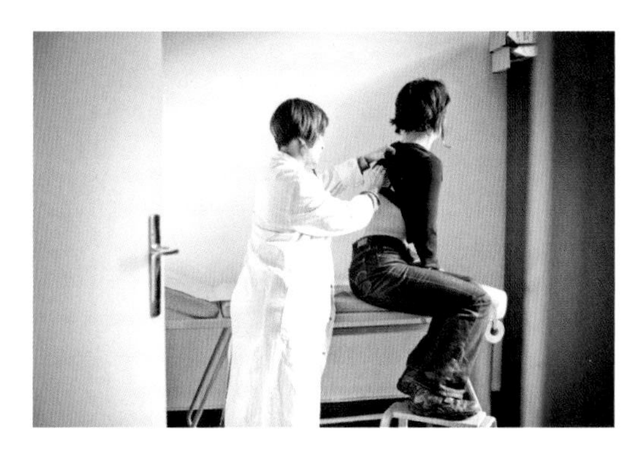

PASTEUR HOSPITAL—ON THE FRONT LINE OF THE HYGIENE BATTLE

In the late 19th century, Pasteurians Émile Roux and Louis Martin often went to hospitals in the Paris public hospital network to treat diphtheria patients. The patients would all be gathered together in one large room, and it was not unusual for them to develop secondary infections after coming into close contact with others. Many died from lung infections, scarlet fever or measles. So when the Pasteur Hospital opened, the two scientists applied Louis Pasteur's theories on hygiene and made sure that all patients were isolated in separate cubicles. The results were staggering: whereas in all other hospitals, 3% of patients were cross-infected by contagious pathogens, in the Pasteur Hospital this figure fell to just 0.3%, or three patients in every thousand. Hospitals treating patients with infectious diseases soon began adapting their wards to implement the isolation system used at the Pasteur Hospital.

Above and Opposite: Consultations at the Institut Pasteur Medical Center.

Vaccination
7
1
2
3

The Institut Pasteur—a Hotbed of Scientific Activity

Biology has become an incredibly complex discipline, and the way research issues are investigated has changed over the past decades. Nowadays it is impossible to identify and demonstrate the relevance of a new concept without bringing experts together and pooling ideas, talents and data. Science no longer operates with a silo mentality in self-contained institutions. With a staff of 2,500 including 1,200 scientists, the Institut Pasteur in Paris clearly reflects this trend.

More than half of its research units are joint structures in partnership with the CNRS or Inserm, and approximately 500 scientists on the Paris campus come from external research organizations (Orex) and are not directly employed by the Institut Pasteur. But they all identify as Pasteurians and are driven by a shared vision of the role science should play. Some Institut Pasteur laboratories work with other institutions, such as the French National Institute for Agricultural Research (INRA), the French National Institute for Computer Science and Applied Mathematics (Inria), and the Paris Public Hospital Network (AP-HP) for clinical research. The Institut Pasteur campus is fully immersed in the French research ecosystem, and it benefits from the expertise of scientists from external organizations, who come with their unique perspective forged by their own scientific culture.

The atrium of the building housing the François Jacob research center at the Institut Pasteur in Paris.

All scientists at the Institut Pasteur are free to pursue partnerships that they believe will be relevant to their research, and these partnerships are often then established by a formal agreement to facilitate dialog and ensure long-term cooperation. In return, the scientists are responsible for ensuring the partnership boosts the Institut Pasteur's reputation at least as much as that of its partner. By way of illustration, the LabEx structures led by Institut Pasteur teams—the product of the "Laboratories of Excellence" program launched by the government a decade ago—are free to cooperate with external laboratories to meet the needs of their research.

The teams also frequently develop partnerships with laboratories in other EU Member States. Scientists at the forefront of their field in areas of strategic importance to the Institut Pasteur are encouraged to apply for EU funding, and they can benefit from support in the application stage, which can be complex. The aim is for the Institut Pasteur to lead the networks set up for EU-funded projects. This policy reflects a "top-down" institutional vision, but one that is firmly rooted in a peer-approved, "bottom-up" strategy.

In recent years, the Institut Pasteur has stepped up cooperation with other institutes in the International Network. But each scientist remains free to draw on personal contacts to establish partnerships with universities, institutions and private laboratories, depending on research requirements.

Since its inception, the Institut Pasteur has always encouraged multidisciplinarity, an approach that spurs innovation, as Louis Pasteur demonstrated in his era. This convergence of scientific disciplines is even more crucial today than it was in the past, both in terms of organization and vision.

The forecourt outside the Scientific Information Center at the Institut Pasteur in Paris.

"The Institut Pasteur campus is resolutely outward-looking, and its partnerships with external researchers and organizations are a valuable source of input to our campus."

Pierre Legrain, Vice-President Development and Grants Office at the Institut Pasteur

Piernicola Spinicelli, head of a collaborative technological R&D project, explains his work on SIM microscopy for imaging live specimens.

GLATIGNY
Mélissa

Training up Future Generations of Scientists

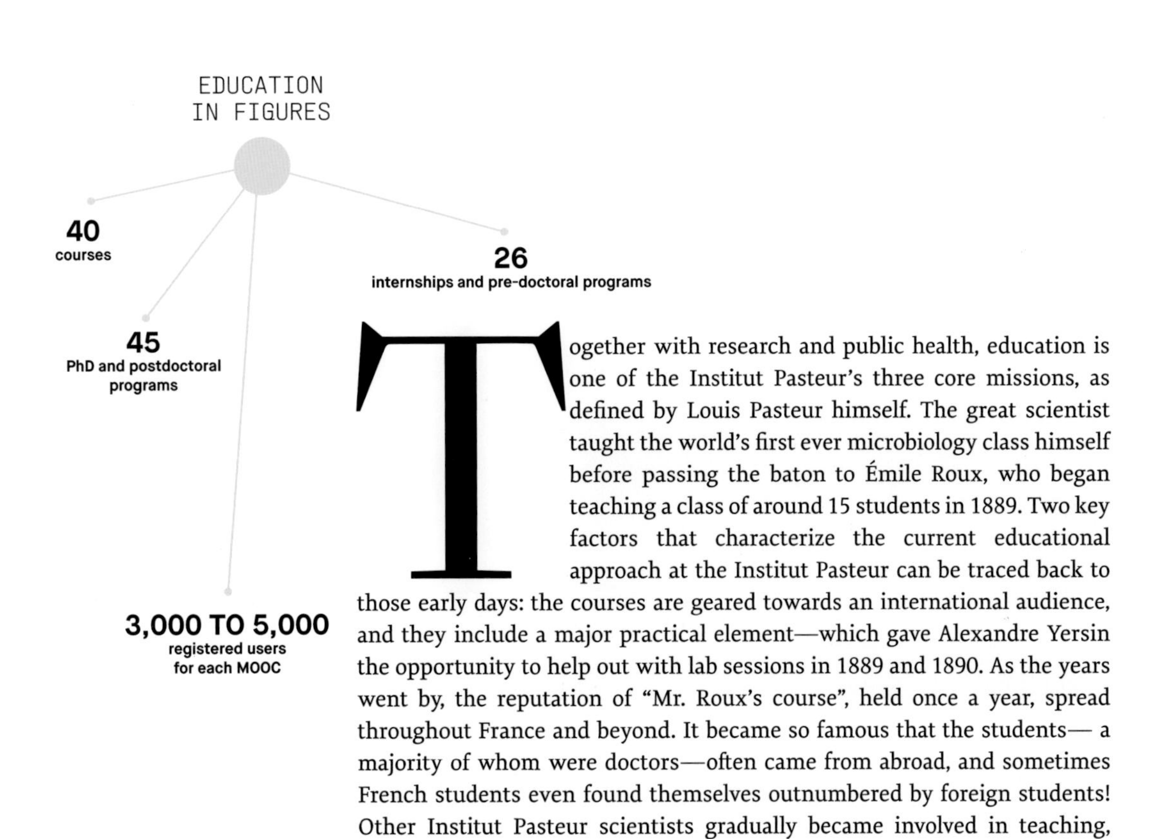

Together with research and public health, education is one of the Institut Pasteur's three core missions, as defined by Louis Pasteur himself. The great scientist taught the world's first ever microbiology class himself before passing the baton to Émile Roux, who began teaching a class of around 15 students in 1889. Two key factors that characterize the current educational approach at the Institut Pasteur can be traced back to those early days: the courses are geared towards an international audience, and they include a major practical element—which gave Alexandre Yersin the opportunity to help out with lab sessions in 1889 and 1890. As the years went by, the reputation of "Mr. Roux's course", held once a year, spread throughout France and beyond. It became so famous that the students— a majority of whom were doctors—often came from abroad, and sometimes French students even found themselves outnumbered by foreign students! Other Institut Pasteur scientists gradually became involved in teaching, including Mechnikov, who took to the task with great enthusiasm when he began in 1890.

Educational activity at the Institut Pasteur was interrupted by the First World War, but it resumed in 1922, with René Legroux and Julien Dumas taking over the practical side. The successor to Émile Roux's famous course

Opposite: Practical sessions at the Institut Pasteur Education Center in 2013.

became known as the *Grand Cours*. Classes were held every afternoon from November to March, and the structure of the course remained virtually unchanged right up to the Second World War. Teaching resumed in 1946-1947 with a broader range of content and the gradual introduction of new disciplines such as microbial physiology, an initiative pioneered by André Lwoff. Given the increasing number of subjects and the sheer volume of content being taught, the curriculum was divided into several separate courses. The first ever immunology course to be taught anywhere in the world began at the Institut Pasteur in 1950.

“As a man, as far back as I can remember I do not believe I have ever met a student without saying to him: Work and persevere!”

Louis Pasteur’s speech at the banquet of the Tercentenary Festival of the University of Edinburgh, April 1884. Works, Vol. VII, p.372.

The microbiology classroom at the Institut Pasteur in Paris, 1910.

A global and interdisciplinary approach

Nowadays the microbiology course—run in partnership with Paris-Descartes, Pierre et Marie Curie and Paris-Diderot universities—includes 60 lectures and workshops held at the Institut Pasteur in Paris or in other International Network member institutes, and attracts more than 1,200 students (Master's and PhD students and professionals) from 70 countries. The sessions are taught by Institut Pasteur researchers—who see education and training as a core part of their role—in cooperation with scientists from all over the world. These scientists are involved in cutting-edge research and are happy to pass on their knowledge of the latest concepts and the best laboratory practices used in science and technological innovation. In addition to the courses run in the Institut Pasteur International Network, Institut Pasteur scientists also regularly contribute to other international teaching programs.

Teaching is geared towards first- and second-year Master's students, PhD students (doctoral school modules), and also professionals (including doctors, pharmacists, veterinarians and researchers), and classes are small, usually with groups of between 14 and 20. Students benefit from highly specialized teaching that can lead to a diploma awarded by the Institut Pasteur, a university diploma, or the Advanced Master's in Public Health awarded by the French National Conservatory of Arts and Trades (CNAM). Many of the courses are taught in partnership with prestigious universities and higher education institutions in France and abroad.

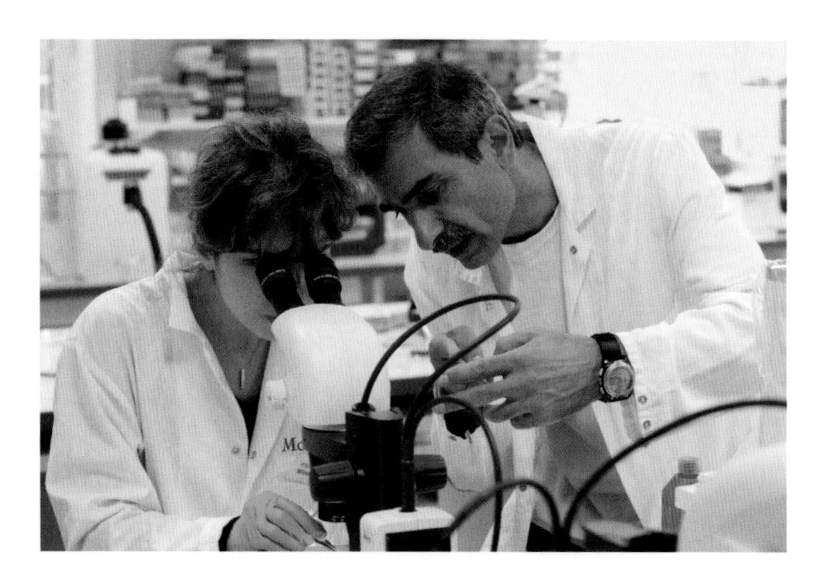

Photo taken at the Education Center during practical sessions for the "Insect Vectors and Transmission of Pathogenic Agents" course in March 2016.

In 2014, the Institut Pasteur began offering online courses (MOOCs), reflecting the latest trends in teaching. The overriding aim of this initiative was to make the courses taught in Paris available to International Network members, and to boost the visibility of the Institut Pasteur's teaching programs. Each MOOC has attracted between 3,000 and 5,000 students and offers an effective way of promoting the instructor-led programs available in Paris. The MOOCs also offer students introductory classes in preparation for the more demanding instructor-led courses taught at one of the International Network member institutes.

An international doctoral program

In 2009, the Institut Pasteur launched its Pasteur-Paris University (PPU) international doctoral program, in cooperation with several leading science universities in Paris, to "attract highly talented and motivated students from all over the world to the Paris campus", explains Susanna Celli, an Institut Pasteur scientist and dean of the PPU program, which trains up "scientists capable of tackling the future biological and medical challenges facing our planet".

This three-year program offers students the opportunity to conduct research in one of the 120 Institut Pasteur laboratories—which are at the forefront in a wide range of topics in all life science disciplines—and to receive comprehensive, up-to-date training. The PPU doctoral program is funded by the Institut Pasteur as well as the European Union, partner institutions, government organizations and private foundations.

Practical sessions at the Institut Pasteur Education Center in 2011.

Practical sessions on blood meal analysis (using ELISA and PCR) during the "Insect Vectors and Transmission of Pathogenic Agents" course in March 2016.

Public Generosity

Since its inception, the Institut Pasteur has been a non-profit foundation officially recognized for charitable status, in accordance with Louis Pasteur's wishes. This special status gives the Institut Pasteur a significant degree of independence, but also means that it depends to a large extent on public generosity.

The story of the Institut Pasteur is also the story of its benefactors, the hundreds of thousands of donors who have supported it over the years. Without them, it would not have been able to build and equip its laboratories, and its scientists would not be able to continue their work, day in, day out, for the benefit of public health. Through their donations and legacies, these benefactors are the beating heart of the Institut Pasteur. It has received almost 4,800 legacies since it was set up, the most newsworthy of which was undoubtedly the major legacy from the Duchess of Windsor in 1986. This generous gift was used to build the Scientific Information Center—complete with an auditorium seating more than 500—which frequently hosts international conferences. Partly through legacies bequeathed over the years, the Institut Pasteur has purchased several apartment buildings in Paris and a wine estate in the Beaujolais region, whose rents contribute to its annual budget. The Institut Pasteur's benefactors come from all walks of life and parts of the world, such as the primary school teacher from Ariège who donated a small amount of her pension every year until she died, and the lone farmer who bequeathed a hectare of vineyards in the Champagne region. And then there are the many artists who donate their works, and the chief executives who involve their companies in fundraising initiatives.

Opposite: Poster for the 2016 Pasteurdon campaign. Pasteurdon is an annual fundraising campaign run by the Institut Pasteur in France.
"Cancer, AIDS, Zika, Alzheimer's, Parkinson's, flu, autism... Together, we can contribute to fighting disease."

Cancers, sida, Zika, Alzheimer, Parkinson, grippe, autisme...

Ensemble, nous avons le don de vaincre les maladies.

Almost a third of the Institut Pasteur's financial resources currently come from individual or corporate donations and from legacies—that's less than the revenue raised from the foundation's activities (particularly royalties from patents), but more than it receives in government grants. Nearly three-quarters of the funds raised through public fundraising are directly allocated to scientific research. To guarantee the transparency of its fundraising operations, the Institut Pasteur Legacies Department is certified to ISO 9001—it was the first foundation in France to be awarded the quality standard by an independent body. Since 2006, the Institut Pasteur has run an annual awareness and fundraising campaign called Pasteurdon, which is supported by a number of celebrities and publicized in the media for several days each year.

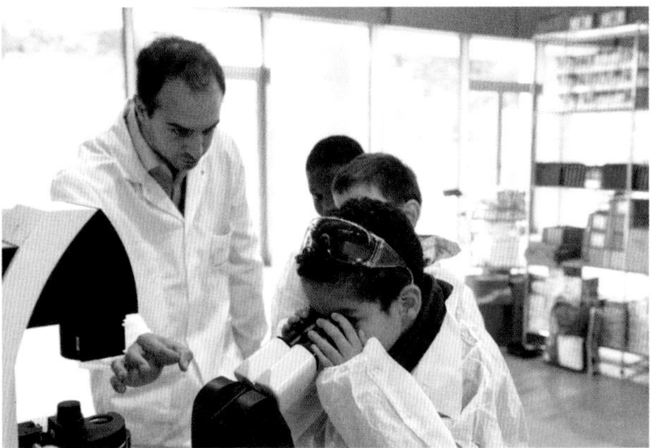

"I have decided to leave my assets to help your scientists. Since I made this decision it's as though a weight has been lifted from my mind, as I know that this money will be used to help advance science."

Mrs M., anonymous donor, 2015

Top left: On October 9, 2016, around 60 runners from the Institut Pasteur took to the starting line of the Paris 20km race. Their sporting attitude and dedication to the cause reflect the shared commitment of Institut Pasteur staff to Pasteurdon.

Top right: A life sciences workshop for primary school pupils during an Institut Pasteur open day in Paris in October 2016.

“Supporting the Institut Pasteur is a logical step for me and enables me to continue the legacy of my father. Helping its scientists in their work is a fascinating human adventure that benefits society as a whole.”

Marina Nahmias, President of the Daniel and Nina Carasso Foundation

Daniel Carasso was a Franco-Spanish businessman and the driving force behind the Spanish food giant Gervais-Danone in the 1970s. He died in Paris in 2009 at the age of 103.

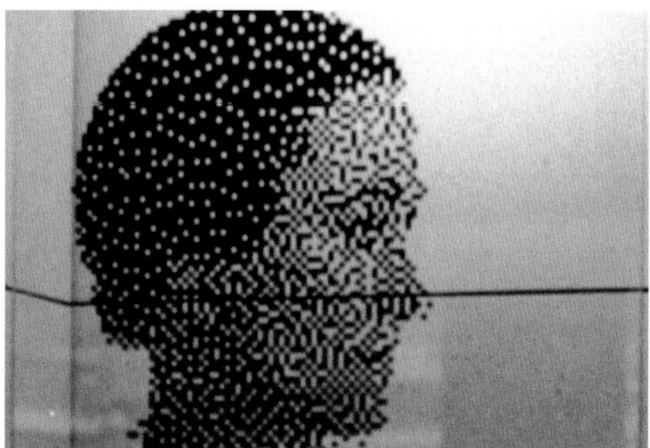

A SUCCESSFUL PUBLIC APPEAL TO BUILD THE INSTITUT PASTEUR

The Institut Pasteur was set up as an institution with official charitable status by decree on June 4, 1887. The foundation took up residence in its first premises the very next year following a successful international public appeal for funds launched by the French Academy of Sciences. It was the first time in France that such a large-scale private initiative had appealed to the public for funds. People of all social strata contributed to the cause, together with a number of eminent figures from abroad— the first donors included a police officer, the Emperor of Brazil and the wealthy owners of the Parisian department store Le Bon Marché. The French Ministries of Public Instruction and Agriculture partly funded the cost of the land and building of the first premises.

The Scientific Information Center (above left) at the Institut Pasteur in Paris was paid for by the legacy of the Duchess of Windsor (above right), who chose the Institut Pasteur as the major beneficiary of her will.

THE INSTITUT PASTEUR'S NOBEL LAUREATES

ALPHONSE LAVERAN (1845-1922),

NOBEL PRIZE 1907

This medical officer, parasitologist and pioneer of tropical medicine discovered the protozoan parasite responsible for malaria in 1880. He was the first to demonstrate that protozoa can cause diseases. He was awarded the Nobel Prize in Physiology or Medicine for his research on the pathogenic role of protozoa.

ILYA MECHNIKOV (1845-1916),

NOBEL PRIZE 1908

This Russian zoologist and biologist gained the Nobel Prize in Medicine for his work on immunity. In 1883, he discovered phagocytes and considered phagocytosis within a wider concept of harmony and disharmony. He used this theory to describe the development of the body and its relationships with microorganisms. He is considered to be the father of cell-mediated immunity (see page 14), as opposed to the humoral immunity observed by Paul Ehrlich. He was jointly awarded the Nobel Prize with Paul Ehrlich in 1908.

JULES BORDET (1870-1961),

NOBEL PRIZE 1919

This leading pioneer of microbiology was awarded the Nobel Prize for his work on the mechanisms of immunity. He shed light on the action mechanism of antibodies and on the role of serum with a set of proteins known as complement, in fighting off infectious agents. His discoveries considerably expanded knowledge of immunology at the time.

CHARLES NICOLLE [1866-1936],

NOBEL PRIZE 1928

He was awarded the Nobel Prize for his work on typhus. He discovered the role of lice in the transmission of the infection in humans. It was then possible to fight outbreaks. Thanks to his research, the disease could be prevented through basic hygiene practices and parasite control.

DANIEL BOVET [1907-1992],

NOBEL PRIZE 1957

Swiss-born Daniel Bovet won the Nobel Prize in Physiology or Medicine for his discoveries relating to synthetic compounds and their effects on blood vessels and skeletal muscles. He discovered the first antihistamine and the first synthetic curare-like agents and pioneered medicinal chemistry.

ANDRÉ LWOFF [1902-1994], FRANÇOIS JACOB [1920-2013] AND JACQUES MONOD [1910-1976],

NOBEL PRIZE 1965

Inspired by André Lwoff, Jacques Monod and François Jacob discovered the first genetic regulation system, for which they coined the term operon. Four years after the publication of this discovery, the three scientists were awarded the Nobel Prize for Medicine. This honor acknowledged the impact of French molecular biology, particularly from the Pasteurian school (see page 19).

FRANÇOISE BARRÉ-SINOUSSI [BORN 1947] AND LUC MONTAGNIER [BORN 1932],

NOBEL PRIZE 2008

These two scientists received the Nobel Prize in Medicine for their work on the discovery of the retrovirus that causes AIDS, carried out at the Institut Pasteur in 1983 (see page 82). Twenty-five years after the AIDS virus was isolated, this prize recognized both their work and the commitment of a whole community in fighting the disease.

Major Institut Pasteur Discoveries to Date

1882 - Work on inflammation and immunity. Discovery of phagocytosis
ILYA MECHNIKOV

1885 - First human rabies vaccination
LOUIS PASTEUR

1888 - Identification of the mechanism of action of the diphtheria toxin
ÉMILE ROUX
ALEXANDRE YERSIN

1894 - Identification of the pathogenic agent responsible for the bubonic plague (*Yersinia pestis*)
ALEXANDRE YERSIN

- Development of diphtheria treatment based on serotherapy
AUGUSTE CHAILLOU
LOUIS MARTIN
ÉMILE ROUX

- Anti-plague serum
ALBERT CALMETTE
AMÉDÉE BORREL

1895 - First plague vaccine using heat-killed bacilli
WALDEMAR HAFFKINE

1896 - Identification of mechanisms underpinning the immune system (especially the involvement of antibodies and role of the complement system)
JULES BORDET

1898 - First cultured Mycoplasma (initially classified as a filterable virus)
EDMOND NOCARD
ÉMILE ROUX

- Role of fleas in the transmission of the plague
PAUL-LOUIS SIMOND

1900 - Observation of the role of trace elements
GABRIEL BERTRAND

1904 - First anti-infectious chemotherapy trials
INSTITUT PASTEUR AND PASTEUR HOSPITAL

1906 - Identification of the bacterium responsible for whooping cough (*Bordetella pertussis*)
JULES BORDET
OCTAVE GENGOU
[I.P. BRABANT]

1907 - Nobel Prize in Medicine awarded to Alphonse Laveran for his work on the role of protozoa in causing diseases. Laveran discovered the protozoan parasite responsible for malaria and called it *Plasmodium*
ALPHONSE LAVERAN

1908 - Nobel Prize in Medicine awarded to Ilya Mechnikov and Paul Ehrlich for their research on immunity
ILYA MECHNIKOV

1909 - Role of lice in the transmission of typhus
CHARLES NICOLLE [I.P. TUNIS]

1910 - Poliomyelitis (polio) is caused by a filterable virus
AMÉDÉE BORREL
CONSTANTI LEVADITI

1917 - Discovery of bacteriophages (independently of F.W. Twort) and invention of the plaque assay method
FÉLIX D'HÉRELLE

1919 - Nobel Prize in Medicine awarded to Jules Bordet for his discoveries relating to immunity
JULES BORDET

1921 - Development of the BCG
ALBERT CALMETTE
CAMILLE GUÉRIN [I.P. LILLE AND I.P. PARIS]

1922-1926 - Development of vaccines linked to the discovery of toxoids
GASTON RAMON
CHRISTIAN ZOELLER

1928 - Nobel Prize in Medicine awarded to Charles Nicolle for his work on typhus
CHARLES NICOLLE

1932 - Development of the yellow fever vaccine
JEAN LAIGRET [I.P. DAKAR]

- Action mechanism of growth factors
ANDRÉ LWOFF

1932-1938 - In lysogenic bacteria, the bacteriophage remains in a non-infectious state
EUGÈNE AND ELIZABETH WOLLMAN

1933 - New plague vaccine
GEORGES GIRARD
JEAN ROBIC

1936 - Discovery of the anti-infectious action of sulfonamides
FEDERICO NITTI
JACQUES AND THÉRÈSE TRÉFOUËL

1937 - Antihistamines and curare-like agents
DANIEL BOVET

1940 - Effective typhus vaccine
HÉLÈNE SPARROW-GERMA
PAUL DURAND

1940-1958 - Work on enzymatic adaptation and enzyme-induced biosynthesis
JACQUES MONOD *ET AL.*

1950-1951 - In lysogenic bacteria, the viral genome is integrated into the host cell like a prophage
ANDRÉ LWOFF

1952 - Development of a method for analyzing antigen mixtures—electrophoretic analysis
PIERRE GRABAR
CURTIS WILLIAMS

1954 - Development of the poliomyelitis vaccine
PIERRE LÉPINE

- Permeases, proteins enabling the specific transport of molecules across bacterial membranes
GEORGES COHEN
JACQUES MONOD *ET AL.*

1954-1958 - Work on bacterial sexuality (conjugation and genetic mapping)
ÉLIE WOLLMAN
FRANÇOIS JACOB

1955 - Work on enzymatic adaptation and enzyme-induced biosynthesis
JACQUES MONOD

1956-1963 - Discovery of immunoglobulin allotypes and antibody idiotypes
JACQUES OUDIN

1957 - Nobel Prize in Medicine awarded to Daniel Bovet for his discoveries relating to synthetic compounds that block the effects of some substances formed and acting in the body, particularly in the blood vessels and striated muscles
DANIEL BOVET

1959-1961 - Regulation of gene expression in bacteria (regulator genes - repressor - operon)
JACQUES MONOD
FRANÇOIS JACOB

1961-1962 - Identification of the messenger RNA and its isolation in different circumstances
JACQUES MONOD
FRANÇOIS JACOB *ET AL.*
FRANÇOIS GROS *ET AL.*

1965 - Nobel Prize in Medicine awarded to François Jacob, André Lwoff and Jacques Monod for their discoveries concerning genetic control of enzyme and virus synthesis
FRANÇOIS JACOB
ANDRÉ LWOFF
JACQUES MONOD

- First enzyme allosteric regulation model
JACQUES MONOD
JEFFRIES WYMAN
JEAN-PIERRE CHANGEUX

1968 - Identification of the activator role of cyclic AMP
AGNÈS ULLMANN
JACQUES MONOD

1968-1970 - New technique for detecting, localizing and quantifying antigens
STRATIS AVRAMEAS

1970 - Isolation of the first neurotransmission receptor
JEAN-PIERRE CHANGEUX

1972-1974 - Identification of protein viral receptors
MAURICE HOFNUNG
MAXIME SCHWARTZ

1974 - Development of bacteriophage vectors in genetic engineering
ALAIN RAMBACH
PIERRE TIOLLAIS

1975 - Cloning of beta-globin complementary DNA
FRANÇOIS ROUGEON
PHILIPPE KOURILSKY

1977 - DNA replication-coupled chromatin assembly
MOSHE YANIV
OLIVIER DANOS

1977-1980 - Multiplicity of human papillomaviruses, specific nature of their pathogenicity, role of one of them in human cancer
GÉRARD ORTH
MICHEL FAVRE
ODILE CROISSANT

1979 - Complete genome sequencing of hepatitis B (first step in the process towards the first human vaccine against this disease)
PIERRE TIOLLAIS AND TEAM

1982 - Complete DNA sequence of the human papillomavirus
MOSHE YANIV
OLIVIER DANOS

1982-1992 - Genetic basis of virulence in *Shigella*, multi-stage molecular process
PHILIPPE SANSONETTI

1983 - Discovery of the AIDS virus
FRANÇOISE BARRÉ-SINOUSSI
JEAN-CLAUDE CHERMANN
LUC MONTAGNIER *ET AL.*

1984 - T4 lymphocytes, LAV human retrovirus receptors
LUC MONTAGNIER *ET AL.*

1985 - Antibodies of human T-lymphotropic virus type I (HTLV-1) in patients with tropical spastic paraparesis
ANTOINE GESSAIN
HUGUES DE THÉ

- Transfer *in vivo* of genetic information from Gram-positive bacteria to Gram-negative bacteria
PATRICE COURVALIN
PATRICK TRIEU-CUOT

1986 - First three-dimensional structure of an antigen-antibody complex
ROBERTO POLJAK *ET AL.*

1987-1992 - Molecular mechanisms of *Listeria monocytogenes* infection
PASCALE COSSART *ET AL.*

1988 - Discovery of a retinoic acid receptor
ANNE DEJEAN

1988-1992 - Cloning of several essential *Helicobacter pylori* genes leading to the development of a rapid diagnostic test
AGNÈS LABIGNE *ET AL.*

1989 - Development of a rapid diagnostic test for tuberculosis
BRIGITTE GICQUEL *ET AL.*

1990 - Targeted replacement of the homeobox gene by the *Escherichia coli* lacZ gene in chimeric mouse embryos
PHILIPPE BRÛLET *ET AL.*

- Cloning of the NF-kB gene
ALAIN ISRAËL
PHILIPPE KOURILSKY

1990-1994 - Discovery of the genetic defect behind acute promyelocytic leukemia (PML-RAR oncoprotein) and explanation of retinoic acid treatment
ANNE DEJEAN

1991 - Mechanisms by which bacterial pathogens enter cells and cross the physical barrier
PASCALE COSSART *ET AL.*

- Development of the Shiga vaccine
PHILIPPE SANSONETTI *ET AL.*

1992-2004 - Role of actin in pathogenic bacteria motility
PASCALE COSSART *ET AL.*

1995 - First pre-clinical trial of the papillomavirus vaccine
FRANÇOISE BREITBURD
GÉRARD ORTH

- Identification of hearing loss genes in humans and the molecular mechanism of hearing
CHRISTINE PETIT *ET AL.*

1996 - Complete sequencing of the eukaryotic genome of *Saccharomyces cerevisiae*
BERNARD DUJON AS PART OF A EUROPEAN PROJECT

1997 - Role of inflammation and anti-inflammation in bacterial strategy for infecting and invading intestinal mucosa
PHILIPPE SANSONETTI *ET AL.*

- Complete sequencing of the *Bacillus subtilis* genome
ANTOINE DANCHIN
FRANCK KUNST
GEORGES RAPPOPORT *ET AL.*

1998 - Complete sequencing of the *Mycobacterium tuberculosis* genome
STEWART COLE *ET AL.*

- First isolation of the virulence gene in *Mycobacterium tuberculosis*
BRIGITTE GICQUEL *ET AL.*

1999 - Identification of a type of nicotinic receptor required for the analgesic action of nicotine
JEAN-PIERRE CHANGEUX *ET AL.*

1999-2002 - Initial identification of genes predisposing to infection by the papillomavirus in humans
GÉRARD ORTH
NICOLAS RAMOZ
MICHEL FAVRE

2000 - Identification of a mechanism of genetic diversity in *Plasmodium*, the malaria agent
ARTUR SCHERF *ET AL.*

2001 - A new endocytosis pathway
ALICE DAUTRY-VARSAT *ET AL.*

- Effective lentiviral vectors derived from HIV-1 for gene therapy
PIERRE CHARNEAU *ET AL.*

- Bacterial genomics and gene regulation: *Listeria monocytogenes et al.*
PASCALE COSSART *ET AL.*

2001-2002 - Comparative genomics: *Listeria monocytogenes et al.*
TEAM LED BY
PASCALE COSSART AND PHILIPPE GLASER

- Comparative genomics: *Mycobacterium leprae*
STEWART COLE *ET AL.*

2003 - Identification of two genes linked to autism
THOMAS BOURGERON *ET AL.*

- Differentiation of new neurons in the olfactory bulb of the adult brain. Evolution, integration and function
PIERRE-MARIE LLEDO *ET AL.*

- Identification of plague bacillus virulence genes
ÉLISABETH CARNIEL *ET AL.*

2004 - Discovery of eukaryotic-type proteins in the *Legionella pneumophila* genome
CARMEN BUCHRIESER

2004-2012 - Identification and regulation of stem cells and muscle progenitor cells during the development and regeneration phase
SHAHRAGIM TAJBAKHSH *ET AL.*

2005 - Transcripts synthesized by RNA polymerase II are subject to a nuclear quality control mechanism involving a new poly(A) polymerase
ALAIN JACQUIER

- Identification of genetic determinants for myogenesis and cardiogenesis
MARGARET BUCKINGHAM *ET AL.*

2006 - Identification of the source of the chikungunya outbreak in the Indian Ocean by sequencing the genomes of viral strains
SYLVAIN BRISSE
ISABELLE SCHUFFENECKER *ET AL.*

2006-2014 - The various developments of old and new DNA strands show the non-equivalence of individual DNA strands
SHAHRAGIM TAJBAKHSH *ET AL.*

2008 - Nobel Prize in Medicine awarded to Françoise Barré-Sinoussi and Luc Montagnier for their work on the discovery of the Human Immunodeficiency Virus (HIV)
FRANÇOISE BARRÉ-SINOUSSI
LUC MONTAGNIER

- A new source of neuron production in the adult brain
PIERRE-MARIE LLEDO
MARIANA ALONSO

2010 - The origins of hematopoietic stem cells
KARIMA KISSA
PHILIPPE HERBOMEL

2011 - The unsuspected immune arsenal in infants
GÉRARD EBERL
SHINICHIRO SAWA

2012 - The dormant state of stem cells which remain viable 17 days post mortem
FABRICE CHRÉTIEN
SHAHRAGIM TAJBAKHSH

2013 - AIDS: 14 adult patients manage to control their HIV infection more than seven years after stopping treatment
ASIER SÁEZ-CIRIÓN *ET AL.*

2014 - Inserm and the Institut Pasteur identify a new variant of the Ebola virus in Guinea
SYLVAIN BAIZE *ET AL.*

2015 - First case of prolonged remission – twelve years – in an HIV-infected child
PIERRE FRANGE
ASIER SÁEZ-CIRIÓN *ET AL.*

2016 - A link is discovered between Zika and microcephaly, with a high risk in the first trimester of pregnancy
SIMON CAUCHEMEZ *ET AL.*

Bibliographical References

Institut Pasteur www.pasteur.fr
World Health Organization (WHO) www.who.int

L. Loison and M. Morange (eds.), *L'Invention de la régulation génétique*, Rue d'Ulm, 2017.
M. Morange, "Les mousquetaires de la nouvelle biologie", *Les Génies de la Science*, no. 10, 2002.
A. Perrot and M. Schwartz, *Le Génie de Pasteur au secours des Poilus*, Odile Jacob, 2016.
A. Perrot and M. Schwartz, *Pasteur et ses lieutenants, Roux, Yersin et les autres*, Odile Jacob, 2013.

THE INSTITUT PASTEUR INTERNATIONAL NETWORK

A. Perrot, M. Schwartz, *Pasteur et ses lieutenants*, Odile Jacob, 2013.
"Les prolongements de l'œuvre pasteurienne dans le tiers monde", *in* M. Morange (ed.), *L'Institut Pasteur, Contributions à son histoire*, La Découverte, 1991.

AT THE FOREFRONT OF RESEARCH

TUBERCULOSIS

R. Brosch, S.V. Gordon, M. Marmiesse *et al.*, "A new evolutionary scenario for the Mycobacterium tuberculosis complex", *Proceedings of the National Academy of Sciences of the United States of America*, 2002, 99(6), p. 3684-3689.
G. Charbonnier (ed.), *La Planète des bactéries*, CIRAD, coll. Les Savoirs Partagés, Montpellier, 2007.
S.T. Cole, R. Brosch, J. Parhill *et al.*, "Deciphering the biology of Mycobacterium tuberculosis from the complete genome sequence", *Nature*, 1998, 39(6685), p. 537-544.
C. La Presle, "Le rôle de l'hôpital de l'Institut Pasteur dans l'application des découvertes fondamentales", *in* M. Morange (ed.), *L'Institut Pasteur, Contributions à son histoire*, La Découverte, 1991.
World Health Organization (WHO), *Global tuberculosis report 2016*.
A. Perrot, M. Schwartz, *Pasteur et ses lieutenants*, Odile Jacob, 2013.
C. Rousseau, M. Winter, E. Pivert *et al.*, "Production of phthiocerol dimycocerosates protects Mycobacterium tuberculosis from the cidal activity of reactive nitrogen intermediates produced by macrophages and modulates the early immune response to infection", *Cellular Microbiology*, 2004, 6(3), p. 277-287.
Jacques Ruffié, Jean-Charles Sournia, *Les épidémies dans l'histoire des hommes*, Flammarion, 1995.

MALARIA

A.C. Allison, "Malaria and sickle-cell anemia", *British Medical Journal*, 1954, 1, p. 290-294.
E. Faway, L. Musset, S. Pelleau *et al.*, "Plasmodium vivax multidrug resistance-1 gene polymorphism in French Guiana", *Malaria Journal*, 2016, DOI 10.1186/s12936-016-1595-9.
J.B.S. Haldane, "The rate of mutation of human genes", Proceedings of the 7th International Conference on Genetics, Hereditas, 1949, sup., p. 267-272.
G. Lambert, *Vérole, cancer & Cie. La société des maladies*, Seuil, 2009.
D. Ménard, M. Khim, J. Beghain *et al.*, "A worldwide-map of Plasmodium falciparum K13-propeller polymorphisms", *New England Journal of Medicine*, 2016, 374, p. 2453-2464.
World Health Organization (WHO), "Malaria", *Fact sheet*, no. 94, April 2017.
Erik Orsenna, *Géopolitique du moustique*, Fayard, 2017.
François Rodhain, *Le Parasite, le moustique, l'homme... et les autres*, Docis, 2015.

AIDS

F. Barré-Sinoussi, J.-C. Chermann, F. Rey *et al.*, "Isolation of a T-lymphotropic retrovirus from a patient at risk of acquired immune deficiency syndrome (AIDS)", *Science*, 1983, 220(4599), p. 868-871.
B. Descours, G. Petitjean, J.L. Lopez-Zaragoza *et al.*, "CD32A is a marker of CD4 T-cell HIV reservoir harbouring replication-competent proviruses", *Nature*, 2017, 543(7646), p. 564-567.

EMERGING INFECTIOUS DISEASES

J.-P. Bado, M. Michel, "Sur les traces du Dr Emile Marchoux, pionnier de l'Institut Pasteur en Afrique noire", *in* M. Morange (ed.), *L'Institut Pasteur, Contributions à son histoire*, La Découverte, 1991.
V.-M. Cao-Lormeau, A. Blake, S. Mons *et al.*, "Guillain-Barre syndrome outbreak associated with Zika virus infection in French Polynesia: a case-control study", *Lancet*, 2016, 387(10027), p. 1531-1539.
S. Cauchemez, M. Besnard, P. Bompard *et al.*, "Association between Zika virus and microcephaly in French Polynesia 2013-2015: a retrospective study", *Lancet*, 2016, 387(10033), p. 2125-2132.
P. Debré, J.-P. Gonzalez, *Vie et mort des épidémies*, Odile Jacob, 2013.
M. D. Grmek, "Le concept de maladie émergente", *History and Philosophy of Life Sciences*, 1993, 15(3), p. 281-296.
F. Keller (rapp.), "Les nouvelles menaces des maladies infectieuses émergentes", Rapport d'information fait au nom de la Délégation sénatoriale à la prospective, no. 638 (2011-2012), July 5, 2012.
M. Morange (ed.), *L'Institut Pasteur, Contributions à son histoire*, La Découverte, 1991.
C. Nicolle, *Destin des maladies infectieuses*, Félix Alcan, 1933.
A. Perrot, M. Schwartz, *Pasteur et ses lieutenants*, Odile Jacob, 2013.
J.-F. Saluzzo, P. Vidal, J.-P. Gonzalez, *Les Virus émergents*, IRD-Editions, 2004.

STEM CELLS

HAL-SHS open archive (Humanities and Social Sciences) https://halshs.archives-ouvertes.fr/
Indian Academy of Sciences www.ias.ac.in
Institut Pasteur www.pasteur.fr
Nobel Prize www.nobelprize.org
The Embryo Project Encyclopedia https://embryo.asu.edu/
A.E. Almada, A.J. Wagers, "Molecular circuitry of stem cell fate in skeletal muscle regeneration, ageing and disease", *Nature Reviews Molecular Cell Biology*, 2016.
A.J. Becker (Dr) *et al.*, "Cytological demonstration of the clonal nature of spleen colonies derived from transplanted mouse marrow cells", *Nature*, February 2, 1963.
A. Chiche *et al.*, "Injury-Induced Senescence Enables In Vivo Reprogramming in Skeletal Muscle", *Cell Stem Cell*, March 2, 2017.
M.-N. Cordonnier, "Des cellules souches dans la rétine", *Pour la science*, news, January 2012.
L. Coulombel, "Des cellules souches pour réparer et régénérer les tissus ?" *Pour la Science*, no. 422, December 2012.
D. Ilic, C. Ogilvie *et al.*, "Concise Review: Human Embryonic Stem Cells-What Have We Done? What Are We Doing? Where Are We Going?", *Stem Cells*, January 2017.
M. Latil *et al.*, "Skeletal muscle stem cells adopt a dormant cell state post mortem and retain regenerative capacity", *Nature Communications*, June 12, 2012.
M. Morange, "Les mousquetaires de la nouvelle biologie", *Les Génies de la Science*, no. 10, 2002.
M. Morange, "François Jacob. June 17 1920–April 19, 2013", *Proceedings of the American Philosophical Society*, vol. 160, no. 2, June 2013.
N. Peyrieras, M. Morange (eds.), *Travaux scientifiques de François Jacob*, Odile Jacob, 2002.

M. Ramalho-Santos, H. Willenbring, "On the Origin of the Term "Stem Cell"", *Cell Stem Cell*, July 2007.
P. Rocheteau *et al.*, "A Subpopulation of Adult Skeletal Muscle Stem Cells Retains All Template DNA Strands after Cell Division", *Cell*, January 20, 2012.
Yu Xin Wang, N.A. Dumont, M.A. Rudnicki, "Muscle stem cells at a glance", *J Cell Sci*, 2014.
B. Wolff, M. Mauget-Faÿsse, J.-A. Sahel, "Quelles pistes pour soigner la DMLA ?", *Pour la Science*, no. 463, May 2016.
"Who really discovered stem cells? The history you need to know", *Knoepfler Lab at UC Davis School of Medicine website*, April 11, 2012.

THE GUT MICROBIOTA

G. Eberl, "Nos bactéries et nous : un équilibre subtil", *Pour la Science*, no. 447, January 2015.
P. Gérard, "Obésité : la flore intestinale mise en cause", *Pour la Science*, no. 447, January 2015.

UNLOCKING THE BRAIN'S MYSTERIES

A. Carleton *et al.*, "Becoming a new neuron in the adult olfactory bulb", *Nature Neuroscience*, May 2003.
K. Deisseroth, "Les neurones sous l'emprise de la lumière", *Pour la Science*, no. 401, March 2011.
L. Katsimpardi *et al.*, "Vascular and neurogenic rejuvenation of the aging mouse brain by young systemic factors", *Science*, May 9, 2014.
G. Lepousez, A. Nissant, P.-M. Lledo, "Adult Neurogenesis and the Future of the Rejuvenating Brain Circuits", *Neuron*, April 22, 2015.
G. Lepousez, A. Nissant, A. K. Bryant, G. Gheusi, C. A. Greer, P.-M. Lledo, "Olfactory learning promotes input-specific synaptic plasticity in adult-born neurons", PNAS, September 23, 2014.
P.-M. Lledo, *Le Cerveau, la machine et l'humain*, Odile Jacob, 2017.
P.-M. Lledo, M. Alonso, M.S. Grubb, "Adult neurogenesis and functional plasticity in neuronal circuits", *Nature Reviews Neuroscience*, March 2006.
M. Morange, F. Wolff, F. Worms (eds.), *L'Homme neuronal, trente ans après. Dialogue avec Jean-Pierre Changeux*, Rue d'Ulm, 2016.
K.A. Sailor *et al.*, "Persistent Structural Plasticity Optimizes Sensory Information Processing in the Olfactory Bulb", *Neuron*, July 20, 2016.
Special issue "Global Neuroscience", *Cell*, November 2, 2016.

AUTISM

T. Bourgeron, "Current knowledge on the genetics of autism and propositions for future research", *Comptes rendus Biologies*, July-August 2016.
T. Bourgeron, M. Leboyer, R. Delorme, "Autisme, la piste génétique", *La Recherche*, no. 426, January 2009.

BIOLOGY AND THE AGE OF BIG DATA

World Health Organization (WHO) www.who.int
The International Genome Sample Resource (IGSR) www.internationalgenome.org/about
What is biotechnology? www.whatisbiotechnology.org/home
R. Baer *et al.*, "DNA sequence and expression of the B95-8 Epstein—Barr virus genome", *Nature*, July 19, 1984.
E. Cassana *et al.*, "Concomitant emergence of the antisense protein gene of HIV-1 and of the pandemic", *PNAS*, October 11, 2016.
J.R. Karr *et al.*, "A Whole-Cell Computational Model Predicts Phenotype from Genotype", *Cell*, July 20, 2012.
U.G. Moritz, D.P. Kraemer, "Spread of yellow fever virus outbreak in Angola and the Democratic Republic of the Congo 2015–16: a modelling study", *The Lancet Infectious Diseases*, March 2017.
P.H. Oliveira *et al.*, "Regulation of genetic flux between bacteria by restriction-modification systems", *PNAS*, May 17, 2016.
H. Quach *et al.*, "Genetic Adaptation and Neandertal Admixture Shaped the Immune System of Human Populations", *Cell*, October 20, 2016.
F. Quetier, P. Wincker, "L'avènement de la métagénomique", *Pour la Science*, no. 81, October-December 2013.
E.P. Rocha, "Using Sex to Cure the Genome", *PLoS Biol.*, March 17, 2016.
P. Tambourin, V. Le Boulc'h, "La génomique à grande échelle", *Pour la Science*, no. 81, October-December 2013.
M. Touchon, E.P. Rocha, "Coevolution of the Organization and Structure of Prokaryotic Genomes", *Cold Spring Harb Perspect Biol.*, January 4, 2016.
R.A. Urbanowicz *et al.*, "Human Adaptation of Ebola Virus during the West African Outbreak", *Cell*, November 3, 2016.

"La sélection naturelle a avantagé une réponse inflammatoire réduite", Entretien avec Lluis Quintana-Murci, *La Recherche*, no. 521, pp. 4-8, March 2017.

BEYOND RESEARCH

EDUCATION

M. Faure, "Cent années d'enseignement à l'Institut Pasteur", *in* M. Morange (ed.), *L'Institut Pasteur, Contributions à son histoire*, La Découverte, 1991.

PUBLIC GENEROSITY

M.-H. Marchand, J. Méry, A. Perrot, *Histoire des dons et legs à l'Institut Pasteur*. Institut Pasteur Brochure, 1987 (updated 2001, 2015 and 2017).

Acknowledgments

The Institut Pasteur would like to thank all those who agreed to be interviewed for or mentioned in this book for helping to paint a picture of some of the key achievements of the Institut Pasteur and its International Network. These insights into today's leading research topics—while not exhaustive—showcase the variety of themes being explored by our "Pasteurians".

Thanks in particular to Mariana Alonso, Rogerio Amino, Philip Avner, Sylvain Baize, Jean-Christophe Barale, Françoise Barré-Sinoussi, Philippe Bastin, Sylvie Bay, Thomas Bourgeron, Hervé Bourhy, Sébastien Boyer, Sylvain Brisse, Roland Brosch, Henri Buc, Carmen Buchrieser, Margaret Buckingham, Élisabeth Carniel, Simon Cauchemez, Susanna Celli, Jean-Pierre Changeux, Pierre Charneau, Frédérique Chegaray, Jean-Claude Chermann, Chetan Chitnis, Fabrice Chrétien, Stewart T. Cole, Paul-Henri Consigny, Pierre-Jean Corringer, Pascale Cossart, Alice Dautry, Laurent Dacheux, Laurent Debarbieux, Anne Dejean, Christiane Demeure, Nicolas Dray, Bernard Dujon, Myrielle Dupont-Rouzeyrol, Gérard Eberl, Anna-Bella Failloux, Arnaud Fontanet, Didier Fontenille, Olivier Gascuel, Odile Gelpi, Christiane Gerke, Antoine Gessain, Gilles Gheusi, Brigitte Gicquel, Tamara Giles-Vernick, Philippe Glaser, Frédéric Grosjean, Jean-Louis Guénet, Chantal Henry, Philippe Herbomel, Alain Israël, Alain Jacquier, Marc Jouan, Lida Katsimpardi, Karima Kissa, Monique Lafon, Claude Leclerc, Pierre Legrain, Gabriel Lepousez, Han Li, Pierre-Marie Lledo, Daniel Louvard, Laleh Majlessi, Jean-Claude Manuguerra, Uwe Maskos, Salaheddine Mécheri, Didier Ménard, Robert Ménard, Luc Montagnier, Michel Morange, Hugo Mouquet, Laurence Mulard, Lise Musset, Pablo Navarro Gil, Jean-François Nicolas, Michael Nilges, Gérard Orth, Perrine Parize, Christine Petit, Armelle Phalipon, Lluis Quintana-Murci, Félix Rey, Eduardo Rocha, Asier Sáez-Cirión, Anavaj Sakuntabhai, Monica Sala, Philippe Sansonetti, Artur Scherf, Maxime Schwartz, Olivier Schwartz, Etienne Simon-Lorière, Pierre Sonigo, Shahragim Tajbakhsh, Frédéric Tangy, Pierre Tiollais, Noël Tordo, Roberto Toro, Maria Van Kerkhove, Marco Vignuzzi, Simon Wain-Hobson, Mary Weiss, and Moshe Yaniv.

Our sincere thanks also go to Christian Vigouroux, Chairman of the Institut Pasteur Board of Directors, Christian Bréchot, President, Jean-François Chambon, Vice-President of Communications and Fundraising and editor of this book, and to Michel Morange, a former Institut Pasteur biologist, historian and philosopher, and the book's editorial adviser. Their enthusiasm for the project and enjoyment in sharing it are what made this book possible.

Lastly, we would like to thank writers Marie-Neige Cordonnier, Émilie Gillet and Gérard Lambert, and the Institut Pasteur teams for overseeing the production of the book, Caroline Etivant, Aurélien Coustillac, Michael Davy, Daniel Demellier and Sandra Legout.

This book is a reflection of what Louis Pasteur once said of scientific discovery:

"[It] is never the work of a single person. Each of those who collaborated in it has contributed many sleepless nights."

Works, Vol. VII, p.163, 164. [Notes on the history of scientific discoveries written in 1858.]

Photography Credits

Cover photo: © Institut Pasteur/Marie-Christine Prévost – Ultrastructural Microscopy Platform/Nathalie Sol-Foulon and Olivier Schwartz, Virus and Immunity Unit/Colorization by Jean-Marc Panaud. Backcover : © Institut Pasteur - Institut Pasteur/François Gardy

P. 8: © Institut Pasteur–Musée Pasteur/François Gardy; p. 10-17, 38, 39, 47, 62, 64 (right), 94, 126, 192, 200: © Institut Pasteur–Musée Pasteur; p. 19, 23, 201 (3rd from top): © Institut Pasteur/Archives of Jacques Monod; p. 20, 49, 75: © Institut Pasteur/Antoinette Ryter; p. 22: © Institut Pasteur/Charles Dauguet/Colorization by Jean-Marc Panaud; p. 25, 58, 113, 133 (bottom), 163, 165, 177, 179, 183-188, 190: © William Beaucardet; p. 27, 33, 86, 117 (top), 118, 128, 150, 160, 168, 189, 193, 195, 199 (left):© Institut Pasteur/François Gardy; p. 29: © Institut Pasteur/Dynamics of Host-Pathogen Interactions; p. 30: © Institut Pasteur/Thérèse Couderc and Marc Lecuit, Biology of Infection Unit/Inserm U1117; p. 37: © Institut Pasteur in Laos; p. 41 (left), 103: © Institut Pasteur in Bangui; p. 41 (right): © Institut Pasteur/Krees Raharison; p. 42: © Institut Pasteur/Dominique Tardy; p. 43: © Institut Pasteur International Network (top), © Institut Pasteur in Laos/Micka Perier (bottom); p. 48: © Institut Pasteur/CNMR archives; p. 50, 55: © Institut Pasteur/Stéphanie Guadagnini – Ultrastructural Microscopy Platform and Mary Jackson, Mycobacterial Genetics Unit; p. 51: © Institut Pasteur/Ludovic Tailleux; p. 52: © Institut Pasteur/Roland Brosch (top), © Institut Pasteur/Giulia Manina (bottom); p. 56: © Institut Pasteur/H. David; p. 60: © Institut Pasteur/Patrice Courvalin; p. 61: © Bruno Deméocq/Total Foundation; p. 64 (left): © Institut Pasteur/Jean-Marc Panaud; p. 67: © Institut Pasteur/Serge Bonnefoy and Monica Diez Silva (top), © Institut Pasteur in French Guiana/Pascal Gaborit (bottom); p. 68, 71-73, 76: © Ronan Liétar/Institut Pasteur in French Guiana; p. 69, 83 (top), 95 (top), 99, 101 (right), 104-106, 113, 114 (left), 158, 166, 201 (2nd from top): © Institut Pasteur; p. 77: © Institut Pasteur/Biology of Host-Parasite Interactions; p. 79, 83 (bottom), 97: © Institut Pasteur/Charles Dauguet; p. 80, 100, 180, 119 (right): © Institut Pasteur/Pierre Gounon; p. 81, 85: © Institut Pasteur/Marie-Christine Prévost – Ultrastructural Microscopy Platform/Nathalie Sol-Foulon and Olivier Schwartz, Virus and Immunity Unit/Colorization by Jean-Marc Panaud; p. 87: © Institut Pasteur/Diagnostics Pasteur; p. 88: © Institut Pasteur/Archives of Odile Croissant; p. 90: © Institut Pasteur/Pierre-Olivier Vidalain and Frédéric Tangy; p. 91: © Institut Pasteur/Perrine Bomme – Ultrapole, Lise Chauveau and Olivier Schwartz, Virus and Immunity Unit/Colorization by Jean-Marc Panaud; p. 93, 201 (top): © Institut Pasteur – Institut Pasteur archives; p. 95 (bottom): © Institut Pasteur/Nicolas Huot; p. 96: © Institut Pasteur/Archives of Alexandre Yersin; p. 98: © Institut Pasteur/Michel Huerre; p. 101 (left): © Institut Pasteur/Hervé Zeller, CNR for Arboviruses and Viral Hemorrhagic Fevers; p. 108: © Institut Pasteur/Nicolas Dray and Laure Mancini; p. 110: © Institut Pasteur/François Jacob (left), © Institut Pasteur/Jean Gaillard and Maurice Lesourd (right); p. 111: © Institut Pasteur/Nadège Cayet, Ultrastructural Microscopy Platform; p. 112: © Institut Pasteur/Jérôme Gros; p. 114 (right): © Institut Pasteur/Jean-François Nicolas; p. 115: © Institut Pasteur/Human Histopathology and Animal Models Unit with the Ultrastructural Microscopy Platform/Colorization by Jean-Marc Panaud; p. 117: © Institut Pasteur/Édith Dériaud and Alexandre Bénéchet with the Dynamic Imaging Platform (bottom, left); © Institut Pasteur/Nicolas Dray (bottom, right); p. 119: © Lymphocyte Development Unit/Photography department; p. 120: © Institut Pasteur/François Schweisguth; p. 121: © Institut Pasteur/Odile Kellermann; p. 122: © Institut Pasteur/Alexandra Tachtsidi; p. 125: © Institut Pasteur/Maurice Lesourd; p. 127, 129, 133 (top), 149: © Institut Pasteur/Gérard Eberl; p. 130: © Institut Pasteur/Perrine Bomme – Ultrastructural Microscopy Platform and Maud Demarque, Nuclear Organization and Oncogenesis Unit; p. 131: © Institut Pasteur/Chantal Ecobichon with the Ultrapole/Colorization by Jean-Marc Panaud; p. 134: © Institut Pasteur/Julie Tomas and Céline Mulet; p. 135: © Institut Pasteur/Benoît Briard and Anne Beauvais, Aspergillus Unit/Perrine Bomme – Ultrastructural Microscopy Platform/Colorization by Jean-Marc Panaud; p. 137: © Institut Pasteur/Karine Gousset and Chiara Zurzolo, Membrane Traffic and Pathogenesis Unit; Christine Schmitt, Ultrastructural Microscopy Platform/Colorization by Jean-Marc Panaud; p. 138, 153: © Institut Pasteur/Pierre-Marie Lledo; p. 140: © Institut Pasteur/Françoise Lazarini, Marie-Madeleine Gabellec and Pierre-Marie Lledo; p. 142, 157: © Institut Pasteur/Valérie Zeitoun; p. 143: © Institut Pasteur/Pierre-Marie Lledo, Perception and Memory Unit; p. 144: © Institut Pasteur/Marie Guibert (left), © INSERM (right); p. 145: © Institut Pasteur/Jean-Pierre Changeux; p. 146, 159: © Institut Pasteur/Roberto Toro; p. 147: © Institut Pasteur/Pierre-Jean Corringer, Marc Delarue, Hughes Nury and Nicolas Bocquet; p. 152: © Institut Pasteur/Fabrice Hyber – Organoïde production; p. 155: © Institut Pasteur/Thomas Bourgeron; p. 169: © Institut Pasteur/Simon Cauchemez; p. 170: © CERMES; p. 171: © Jérôme Bon; p. 173: © M. Jung (CNRS), O. Gascuel (CNRS/IP), M. Peeters (IRD); p. 181: © Institut Pasteur/Meriadeg Le Gouil and Gilberte Coralie (top), © Institut Pasteur/Meriadeg Le Gouil (bottom); p. 182: © Institut Pasteur/Archives of Jacques Bourichon; p. 194: © Institut Pasteur/Béatrice de Cougny; p. 197: © Maxyma/Institut Pasteur; p. 198: © Institut Pasteur/Jean-François Charles; p. 201 (bottom): © Institut Pasteur/Christophe Soubert.

Editorial adviser to Éditions de La Martinière: Michel Morange

Written by:
Marie-Neige Cordonnier
Emilie Gillet
Gérard Lambert

Translation from original French text: id2m

Graphic design and artwork: Grégory Bricout

Distributed in 2018 by Abrams, an imprint of Abrams

ISBN 978-1-4197-3090-0

Color separation: IGS-CP (16)

Printing in 2017 in Portugal

10 9 8 7 6 5 4 3 2 1

With the support of

www.pasteur.fr

ABRAMS The Art of Books
115 West 18th Street, New York, NY 10011
abramsbooks.com